Water Requirements for Irrigation and the Environment

The authors can be reached at the following address:

International Institute for
Geo-Information Science and Earth
Observation, ITC
P.O.Box 6, 7500 AA Enschede
The Netherlands
marinusgbos@cs.com
www.bos-water.nl

Alterra, Wageningen UR
Droevendaalsesteeg 3, 6708 BP
Wageningen, The Netherlands
rob.kselik@wur.nl

University of Idaho
Research and Extension Center
Kimberly, U.S.A
rallen@kimberly.uidaho.edu

International Water Management
Institute, IWMI
P.O.Box 2075, Colombo,
Sri Lanka
d.molden@cgiar.org
www.iwmi.cgiar.org

Previous versions of CRIWAR:

Version 1.* Bos, M.G. 1988, Crop irrigation water requirements, ILRI, Wageningen (limited distribution)

Version 2.* Bos M.G., J. Vos, and R.A. Feddes. 1996. CRIWAR 2.0: A Simulation Model on Crop Irrigation Water Requirements. ILRI, Wageningen

Marinus G. Bos • Rob A.L. Kselik
Richard G. Allen • David J. Molden

Water Requirements for Irrigation and the Environment

Springer

Marinus G. Bos
International Institute for Geo-Information Science and Earth Observation
ITC

Richard G. Allen
University of Idaho
Research and Extension Center

Rob A.L. Kselik
International Institute for Land Reclamation and Improvement
Alterra-ILRI

David J. Molden
International Water Management Institute
IWMI

ISBN: 978-1-4020-8947-3 e-ISBN: 978-1-4020-8948-0

Library of Congress Control Number: 2008933942

Printed on acid-free paper

9 8 7 6 5 4 3 2 1

springer.com

Abstract M.G. Bos, R.A.L. Kselik, R.G. Allen and D.J. Molden 2008. *Water Requirements for Irrigation and the Environment.* Springer, Dordrecht, ISBN 978-1-4020-8947-3

Irrigated agriculture produces about 40% of all food and fibre on about 16% of all cropped land. As such, irrigated agriculture is a productive user of resources; both in terms of yield per cropped area and in yield per volume of water consumed. Many irrigation projects, however, use (divert or withdraw) much more water than consumed by the crop. The non-consumed fraction of the water causes a variety of undesirable effects ranging from water-logging and salinity within the irrigated area to downstream water pollution.

This book discusses all components of the water balance of an irrigated area; evapotranspiration (Chapter 2), effective precipitation (Chapter 3) and capillary rise from the groundwater table (Chapter 4). Chapter 5 then combines all components into a water management strategy that balances actual evapotranspiration (and thus crop yield) with the groundwater balance of the irrigated area (for a sustainable environment). Chapter 6 presents CRIWAR 3.0, being a simulation program which transfers the estimated evapotranspiration of the cropped area into the water requirements of an irrigated area.

The computer program presented in this publication can accommodate a wide variety of cropping patterns as well as many different input and output units. This version greatly expands upon the capabilities of previously published programs.

Keywords Water management; irrigation; groundwater; drainage; environment; water balance; crop production.

Preface

Each day, the continuing growth of world population places new demands on our water resources. More water is needed for all the processes of life: food production, municipal supply, industrial water use, power generation, navigation, recreation, etc. At the same time, environmental water needs are increasingly being recognized, limiting the sources of new water and further increasing the competition for available supplies.

Improved management of our water resources is needed to ensure the equitable distribution of water to competing users. There are especially significant opportunities for conservation and more effective water use by the world's largest user: agriculture. Accurate delivery of the necessary amounts of water at the correct times can both conserve water and improve the quantity and quality of agricultural products. Thus, the method to quantify the irrigation water requirement described in this manual has a key role to play as we address the future water, food, and fibre needs of our world.

This manual gives additional information on capillary rise as a source of water and on the method by which groundwater table management can be used to reduce the surface water requirement during the peak season. This groundwater table management also reduces the need for artificial drainage and thus reduces the negative effect of drainage effluent on the downstream ecosystem.

In addition, the CRIWAR software can be a helpful tool in the management of operational irrigation projects with frequent changing cropping patterns and in the performance assessment of irrigation and drainage.

The range of potential applications for this book and related software is unlimited. We hope that this book will contribute to the effective management of one of the earth's most widely needed, used, and visible natural resources: water.

Marinus G. Bos, Enschede, The Netherlands
Rob A.L. Kselik, Wageningen, The Netherlands
Richard G. Allen, Kimberly, Idaho, USA
David J. Molden, Colombo, Sri Lanka

Contents

List of Symbols

C	height of capillary rise (m)
C_d	denominator constant that changes with reference type and calculation time step (s m^{-1})
C_n	numerator constant that changes with reference type and calculation time step (K mm s^3 Mg^{-1} day^{-1} or K mm s^3 Mg^{-1} h^{-1})
c	dimensionless adjustment factor
c_p	specific heat of dry air at constant pressure (J/kg K)
D_M	day of the month (1–31)
$D_{e,j-1}$	cumulative depletion from the soil surface layer at the end of day j - 1 (the previous day in mm)
d_r	inverse relative distance factor (squared) for the earth-sun (dimensionless)
DF	depleted fraction (dimensionless)
$DP_{ei,j}$	deep percolation in mm from the f_{ew} fraction of the soil surface layer on day j if soil water content exceeds field capacity
E_a	isothermal evaporation rate (kg/m^2 s)
E_j	evaporation in mm on day j (i.e., $E_j = K_e\ ET_0$)
E_0	open water evaporation rate (kg/m^2 s)
e_a	mean actual vapor pressure at 1.5–2.5 m height (kPa), Δ is slope of the saturation vapor pressure versus temperature curve (kPa °C^{-1})
e^o	saturation vapor pressure function
e_s	saturation vapor pressure at 1.5–2.5 m height (kPa), calculated for daily time steps as the average of saturation vapor pressure at maximum and minimum air temperature
e_z	prevailing vapour pressure in the external air, measured at the same height as T_z (k/Pa)
ET_0	standardized reference ET for a 12 cm tall, cool season grass in mm day^{-1} for daily time steps or mm h^{-1} for hourly time steps
ET_a	actual evapotranspiration (mm/day)

$ET_{a,gross}$	sum of the actual evapotranspiration from the (irrigated) cropped area and all fallow (non-cropped) area within the command area served by the irrigation system
$ET_{a,non.ir}$	actual evapotranspiration from all fallow (non-irrigated) area within the command area
ET_L	target landscape ET (in mm day^{-1}, mm month^{-1}, or mm year^{-1})
ET_p	potential; evapotranspiration (mm/day)
F	actual retention (mm)
F↓	downward force (N)
f	correction factor which depends on the depth of the irrigation water application per turn (dimensionless)
$f_{cd\ \beta > 0.3}$	cloudiness function for the time period prior to when β falls below 0.3 radians during afternoon or evening (dimensionless)
f_{cd}	cloudiness function [dimensionless] and limited to $0.05 \leq f_{cd} \leq 1.0$
f_{ew}	fraction of the soil that is both exposed to solar radiation and that is wetted
f(u)	wind function; $f(u) = 1 + 0.864u_2$
G	soil heat flux density at the soil surface in MJ m^{-2} day^{-1} for daily time steps or MJ m^{-2} h^{-1} for hourly time steps
G_{day}	daily (24-h) soil heat flux density (MJ m^{-2} day^{-1})
G_{sc}	solar constant (4.92 MJ m^{-2} h^{-1})
g	acceleration due to gravity ($g = 9.81$ m/s^2)
H	flux density of sensible heat into the air (W/m^2)
h	(hydraulic) head (m)
I_a	initial abstraction (mm)
I_j	irrigation depth in mm on day j that infiltrates the soil
J	number of the day in the year between 1 (1 January) and 365 or 366 (31 December)
K	hydraulic conductivity as a function of h (m/day)
K_c	crop coefficient
K_{cb}	basal crop coefficient [between 0 to 1.4]
K_e	soil water evaporation coefficient [between 0 to 1.4]
K_r	evaporation reduction coefficient
L_z	longitude of the center of the local time zone (expressed as positive degrees west of Greenwich, England)
M	number of the month (1–12)
NDVI	Normalized Difference Vegetation Index
P	total precipitation (mm/day or mm/month)
P_j	precipitation in mm on the soil surface on day j
P_e	effective precipitation (mm/day or mm/month)
p_a	atmospheric pressure (kPa)

r_a	aerodynamic diffusion resistance, assumed to be the same for heat and water vapour (s/m)
R_n	calculated net radiation at the crop surface in MJ m^{-2} day^{-1} for daily time steps or MJ m^{-2} h^{-1} for hourly time steps
R_{ns}	net short-wave radiation, (MJ m^{-2} day^{-1} or MJ m^{-2} h^{-1})
R_{nl}	net outgoing long-wave radiation, (MJ m^{-2} day^{-1} or MJ m^{-2} h^{-1})
R_s	incoming solar radiation (MJ m^{-2} day^{-1} or MJ m^{-2} h^{-1})
T	mean daily or hourly air temperature at 1.5–2.5 m height (°C)
T_{av}	daily average air temperature (°C); $T_{av} = (T_{max} + T_{min})/2$
$T_{ei,\,j}$	day j
T_{min}	minimum air temperature, °C
T_{max}	maximum air temperature, °C
$T_{K\,min}$	minimum absolute temperature during the 24-hour period (K). K = °C + 273.15
$T_{K\,max}$	maximum absolute temperature during the 24-hour period (K)
$T_{K\,hr}$	mean absolute temperature during the hourly period (K)
T_p	statistical value (dimensionless)
T_s	temperature at the evaporating (water) surface (°C)
T_z	air temperature at a height z above the surface (°C)
t	standard clock time at the midpoint of the period in h (after correcting time for any daylight savings shift)
TD	difference between mean daily maximum and minimum temperature (°C); $TD = (T_{max} - T_{min})$
TEW	total evaporable water (mm)
Y	number of the year (for example 1996 or 96)
R_A	extraterrestrial radiation (MJ/m^2 per day) or (MJ m^{-2} h^{-1})
REW	readily evaporable water (mm)
RO_j	runoff in mm from the soil surface on day j
R_a	field application ratio (dimensionless)
S	maximum potential difference between precipitation and runoff beginning at the time precipitation starts (also named maximum retention) (mm)
Q	accumulated runoff depth (mm)
P	accumulated precipitation depth (mm)
p	pressure energy per unit of volume (Pa)
q	vertical flow rate per unit area (m/day)
s	standard deviation
u_2	wind speed at 2.0 m above ground surface (m/s)
V_c	total volume of surface water diverted from the water source (river, reservoir) into the irrigation system
V_d	total volume of irrigation water supplied to the inlets of the distribution system

V_f volume of irrigation water delivered to the fields during the period under consideration (m^3/period)

V_{grw} total volume of groundwater pumped into the conveyance system

V_m volume of irrigation water needed, and made available, to avoid undesirable stress in the crops throughout the growing cycle (m^3/period)

$V_{non\text{-}ir}$ total volume of water supplied for non-irrigation purposes. In most irrigation systems this volume is negligible with respect to V_d

Z_e effective depth of the surface soil subject to drying to 0.5 θ_{WP} by way of evaporation (m)

z elevation head, being positive in the upward direction (m)

z station elevation above sea level (m)

α albedo, fixed at 0.23 for both daily and hourly time steps for reference ET (dimensionless)

β angle of the sun above the horizon (radians)

γ psychrometric constant (kPa $°C^{-1}$)

Δ slope of the saturation vapor pressure versus temperature curve (kPa $°C^{-1}$)

δ solar declination (radians)

ε ratio of molecular masses of water vapour over dry air (dimensionless)

θ_{FC} soil moisture at field capacity ($m^3\ m^{-3}$)

θ_{WP} soil moisture at wilting point ($m^3\ m^{-3}$)

λ latent heat of vaporization (J/kg)

λE flux density of latent heat into the air (W/m^2)

ρ density of water ($\rho = 1{,}000\ kg/m^3$)

ρ_a density of moist air (kg/m^3)

σ Stefan-Boltzmann constant (4.901×10^{-9} MJ $K^{-4}\ m^{-2}\ day^{-1}$ and 2.042×10^{-10} MJ $K^{-4}\ m^{-2}\ h^{-1}$)

φ station latitude (radians)

ω_1 solar time angle at beginning of period (radians)

ω_2 solar time angle at end of period (radians)

ω_s sunset hour angle (radians)

Chapter 1
Introduction

1.1 Growth of Vegetation

For vegetation to grow, it should transpire sufficient water through the stomata on its leaves. This water is taken from the soil via the roots. The part of the soil from which the roots take water is named the effective root zone (Fig. 1.1). Water also moves into the atmosphere through evaporation from plant surfaces (following precipitation or irrigation) and from the bare soil surface in between the vegetation.[1] Part of the water that evaporates from the bare soil surface originates from precipitation or irrigation. The remaining part rises through capillary action from the groundwater table to the soil surface. The sum of the evaporation and transpiration is known as *EvapoTranspiration* (*ET*). If sufficient water is available to meet the sum of evaporation and transpiration, the *ET* will reach its (maximum) potential value, ET_p. Otherwise, the actual evapotranspiration (ET_a) will be less than ET_p (see Chapters 4 and 5).

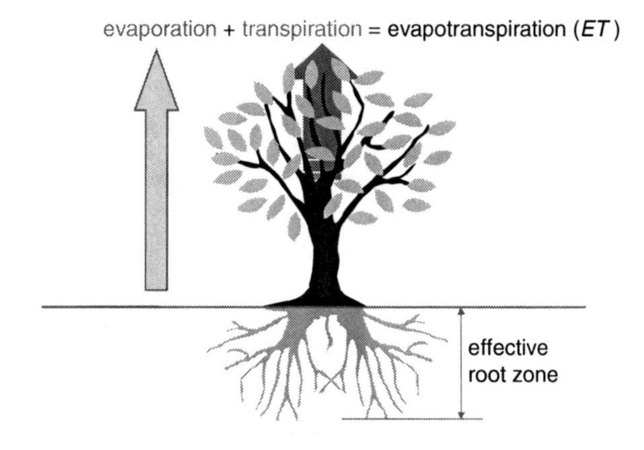

Fig. 1.1 Terminology

[1] Water evaporating directly from the groundwater surface is ignored in this context (also see Chapter 4).

M.G. Bos et al. *Water Requirements for Irrigation and the Environment*,

The potential evapotranspiration, ET_p, is the volume of water required to meet the crop's potential evapotranspiration over the whole growing season, under a given cropping pattern and in a specific climate.

Figure 1.2 shows the mondial distribution of average annual values of the relative evapo-transpiration (ET_a/ET_p). Traditionally, the main areas of food production have been areas with relatively fertile soils, a sufficient supply of water, and favourable climatic conditions. The qualification 'sufficient supply of water' can be re-phrased as: rain-fed agriculture traditionally is practised in areas where the average annual value of ET_a/ET_p is greater than about 0.8, otherwise irrigation was introduced. For some decades, irrigation has also been used as a form of 'insurance' on yield reductions due to dry spells and to control the uniform quality of high value (export) crops.

Because of the increasing demand for agricultural products by a rapidly growing world population, agriculture has expanded horizontally into areas where conditions for production are less favourable. It has also expanded vertically by increasing production per unit area of land through intensification. As a result of this horizontal and vertical expansion, agricultural production has increased considerably. Food and fibres presently are grown on about 1,500 million hectares rain-fed land and 250 million hectares irrigated land. However, the latter 14% of the agricultural area produces 40% of all crops. Hence, irrigation plays a major role in feeding the world.

To meet the growing demand for food and fibre, crop production should increase. However, from a land and water use perspective there are two major constraints:

- ***LAND*** is the traditional constraint. If more crops were needed, more land was reclaimed while the goal was to maximise yield in terms of kg/ha. However, for the last four decades, most suitable areas have already been cropped while urban development infringes on agricultural areas.
- ***WATER*** is the ever more important constraint. Already for 10% of world population (in arid and semi-arid countries) the annually available volume of water

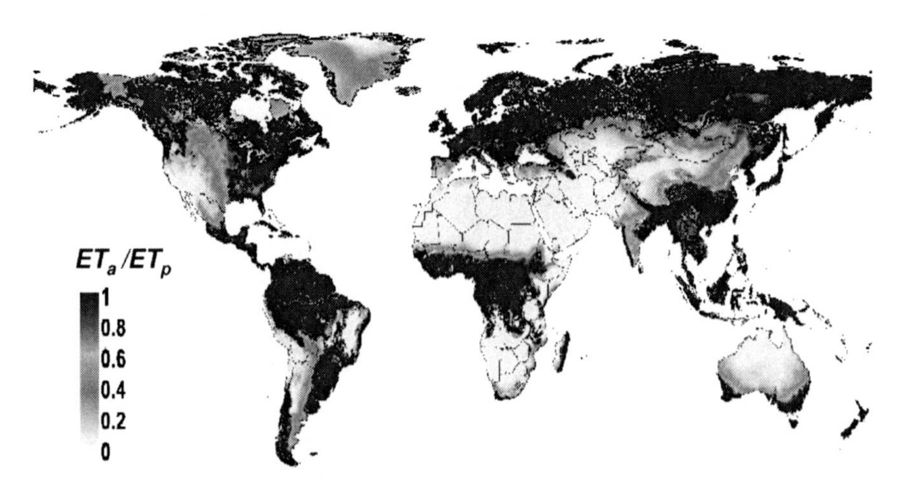

Fig. 1.2 Average annual distribution of the relative evapotranspiration (Adapted from Droogers et al. 2001)

dropped below the critical level of 1,700 m^3 per capita (Fig. 1.3). In such areas, the crop yield in terms of kg/m^3 water becomes increasingly important. Also the quality of water (reuse and disposal) is due to become increasingly important. However, more alarming is that the next group in Fig. 1.3, being 49% of world population, that will pass the water scarcity limit before 2025.

The future challenge is to grow sufficient food on current agricultural land; thus without undue infringement on nature. This should be done in such a way that water use does not damage the environment. Hence, the water balance within the agricultural area should remain stable. To meet this challenge, we follow two tracks which merge into a water use strategy:

Crop production track

This track starts with an estimate of the crop water requirements in order to produce food (and fibre). It then estimates the additional water required to operate an irrigation system. Combining these requirements yields the irrigation water demand of the irrigated area.

Water balance track

This track considers the three major components of the water balance of an irrigated area: actual ET, precipitation, and actual irrigation water supply. These components are matched in such a way that the groundwater table in the area remains stable.

Water use strategy

Merging the irrigation water demand and the actual (planned) irrigation water supply, results in a water use strategy that allows the production of a crop within a stable environment.

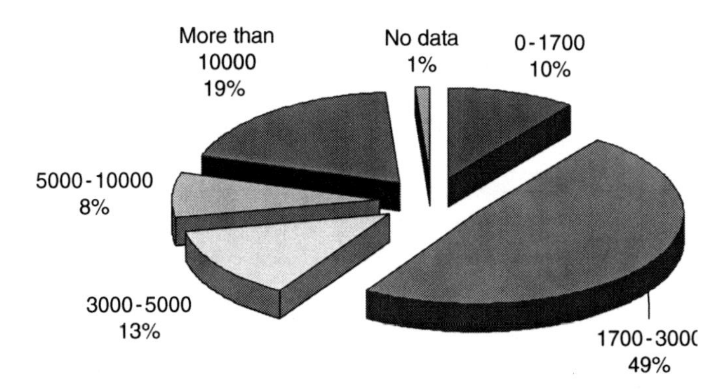

Fig. 1.3 Distribution of world population among economies grouped by annual freshwater resources in cubic metre per capita (World Bank 1999)

1.2 Crop Growth and Evapotranspiration

As mentioned above, the crop transpires water during its growth. With respect to crop water requirement we distinguish four different stages of crop development that are considered for field and vegetable crops (Fig. 1.4):

- The initial growth stage being the germination and early growth stage of the crop. During this stage, the soil surface is not, or is hardly, covered by the crop canopy (ground cover less than 10%). Although transpiration stress can be very harmful during this stage, most water will evaporate from the soil. Hence, during this stage the crop type has little effect on the ET_p-value.
- Crop development stage: lasting from the end of the initial stage until the attainment of effective full ground cover (between 70% and 80%). Please note that this does not mean that the crop has reached its matured height.
- Mid-season stage: lasting from the attainment of effective full ground cover to the start of maturing of the crop. Maturing of the crop may be indicated by leaves discolouring (beans) or leaves falling off (cotton). For some crops, this stage may last until very near harvest (sugar beet) unless irrigation is omitted at late season and a reduction in ET_p is induced to increase yield and/or quality (sugarcane, cotton, some grains). Normally this stage lasts well past the flowering stage of annual crops.
- Late season stage: lasting from the end of the mid-season stage until full maturity or harvest of the crop.

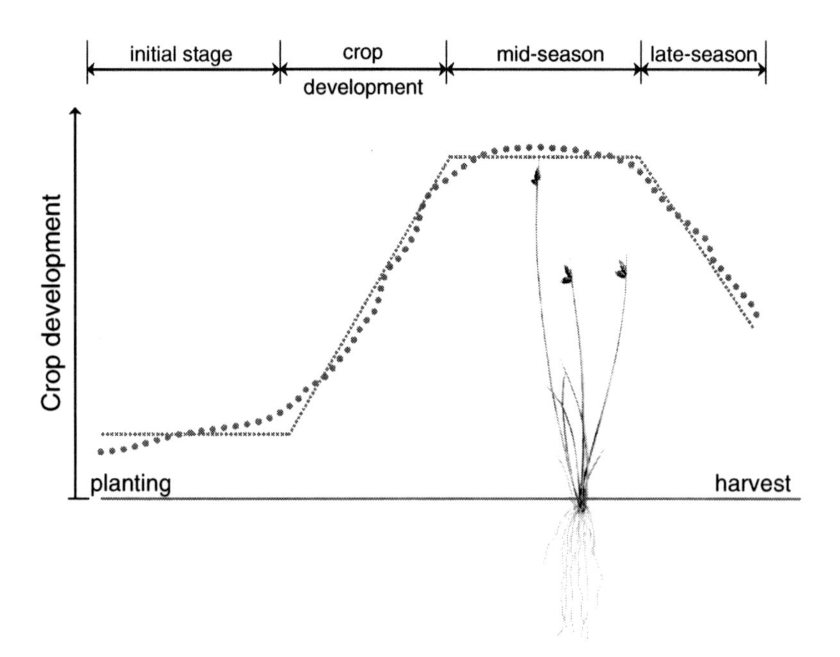

Fig. 1.4 Actual crop development and four schematised growth stages

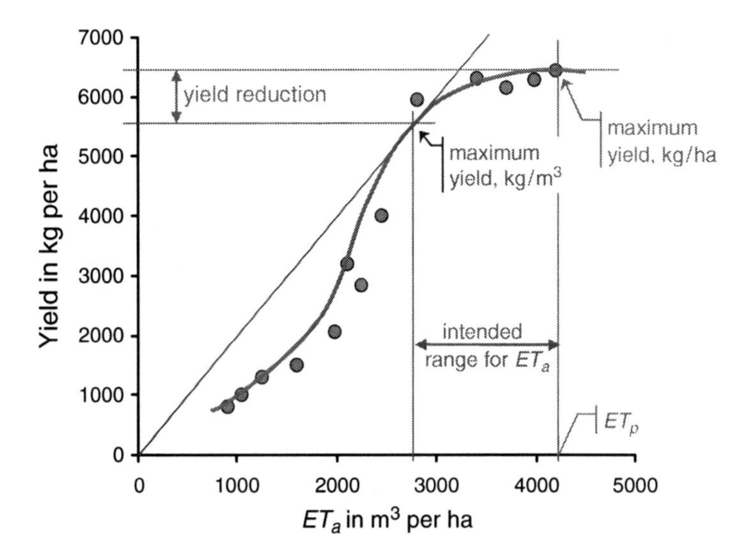

Fig. 1.5 Crop production function showing the cumulative ET_a versus yield for wheat, Central Great Plains, USA (Data points from Hanks et al. 1967)

During each growth stage the ET_a will be equal to ET_p if the crop is not water stressed. The yield of the crop then is maximized in terms of kg/ha as illustrated in the crop production function of Fig. 1.5. Normally, however, the crop will feel some water stress resulting in a lower cumulative ET_a and a lower yield. Depending on the width of the upper (mid-season) part of the crop production function, the ratio ET_a/ET_p can be reduced considerably while crop yield remains high. In the wheat example of Fig. 1.5, the ratio ET_a/ET_p may decrease to 0.67 (being 2,800/4,200) while yield in terms of kg/ha only decreases to 0.87 (being 5,600/6,400) of potential yield. In fact, with this decrease in ET_a the productivity in terms of kg/m^3 will become maximum, which should be the operational target if water is the limiting resource (Bos 1980). For the wheat example of Fig. 1.5, the intended value of ET_a/ET_p should thus be greater than 0.67.

1.3 The Water Balance of an Area

The water balance of a gross command (irrigation) area shows three sources of water: precipitation, groundwater inflow and river (surface) water diversion (Fig. 1.6). Part of all this water evapo-transpires from irrigated crops (fields) and partially from fallow land. This gross evapotranspiration is denoted as:

$$ET_{a,gross} = ET_a + ET_{a,non.ir} \qquad 1.1$$

Where:

$ET_{a,gross}$ = The sum of the actual evapotranspiration from the (irrigated) cropped area and all fallow (non-cropped) area within the command area served by the irrigation system

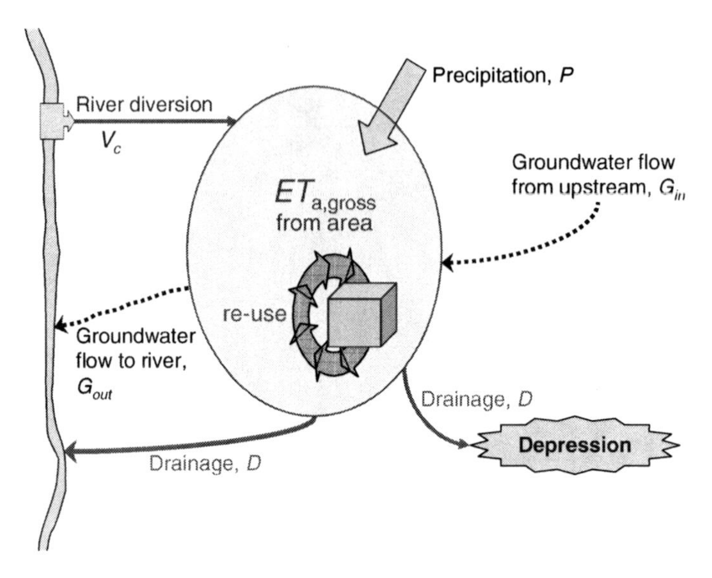

Fig. 1.6 The water balance of a gross command) area with irrigation

ET_a = The actual evapotranspiration from the cropped area within the irrigable area
$ET_{a,non.ir}$ = The actual evapotranspiration from all fallow (non-irrigated) area within the command area

Part of the command area will be permanently non-irrigated (land along canals, roads, villages, etc.). This part often ranges between 5% and 10% of the gross command area. The $ET_{a,non.ir}$ from this area depends on the ground cover (e.g. trees, grass, asphalt, houses). The remainder of the non-irrigated area consists of fields being fallow in between harvest and seeding/planting. For this part we assume that $ET_{a,non.ir}$ equals the evapotranspiration during the initial growth stage (see Section 1.2).

The part of the available water that does not evaporate will flow to downstream areas either via surface streams (drains) or as groundwater. If the summed inflow exceeds the outflow, part of the water will be stored within the irrigated area. This increased storage may cause water logging and salinity. If, on the other hand, the summed outflow exceeds the inflow, the groundwater table will drop. In first instance this will reduce the availability of capillary water to crop growth. With continued mining of groundwater, this water resource will be depleted. To avoid the above problems it is recommended to manage irrigation water (V_c) in such a way that the groundwater table remains stable from year to year.

To avoid the accumulation of salts (in the root zone of the crops) within the irrigated area of Fig. 1.6, about 10–20% (say 15%) of the total inflow ($V_c + P + G_{in}$) should discharge from the area as drainage (D) plus groundwater outflow (G_{out}). In other words; $ET_{a,gross}$ should be less than about 85% of the available water (inflow). Thus, for sustainability:

$$\frac{ET_{a,gross}}{V_c + P + G_{in}} \leq 0.85 \qquad 1.2$$

The groundwater inflow (G_{in}) is usually low in comparison with the other three flow volumes and is difficult to quantify unless a groundwater model of the area is available. If we remove G_{in} from Equation 1.2 it reduces to the depleted fraction, being defined as (Molden 1997; Bastiaanssen et al. 2001; Bos 2004):

$$depleted\ \ fraction = \frac{ET_{a,gross}}{V_c + P} \qquad 1.3$$

The water balance of the gross area can be characterized through this depleted fraction. It relates the actual evapotranspiration from the gross area to the sum of all precipitation on this area plus the surface water inflow, V_c (irrigation water) into the area.[2] The depleted fraction quantifies the surface water balance excluding the drainage component. The water manager can influence the volume of supplied irrigation water, V_c, while this volume in turn influences the actual evapotranspiration (ET_a) from the irrigated fields. Chapter 5 shows how the limits on crop growth, sustainability and water resource use result to the recommendation to manage irrigation in such a way that the depleted fraction ranges between 0.5 and 0.9.

1.4 The Water Balance of an Irrigated Field

The water balance of an irrigated field (Fig. 1.7) should be viewed from three perspectives: a crop growth perspective, a sustainability perspective, while the increasingly scarce water resource should be used efficiently. Hence, a match needs to be found between the following partly conflicting rules:

- To facilitate crop growth, water stress should be limited especially during the first three growth stages (Section 1.2). For the wheat example of Fig. 1.5, this means that the relative evapotranspiration ET_a / ET_p should be greater than about 0.67 within the irrigated fields. The time steps for which this fraction should be quantified vary between 7 to 10 days.
- For sustainable agriculture, the accumulation of chemicals (salt, pesticides, etc.) in the root zone must be avoided. Since all chemicals are transported by water, this means that the annual downward seepage from the root zone must exceed the annual capillary rise into the root zone by some 10–20% (Chapter 4). The accumulation of chemicals can be tolerated during dry months provided that they will be leached during the following wet months.

[2]Groundwater being mined from a deep aquifer should be added to the irrigation water supply.

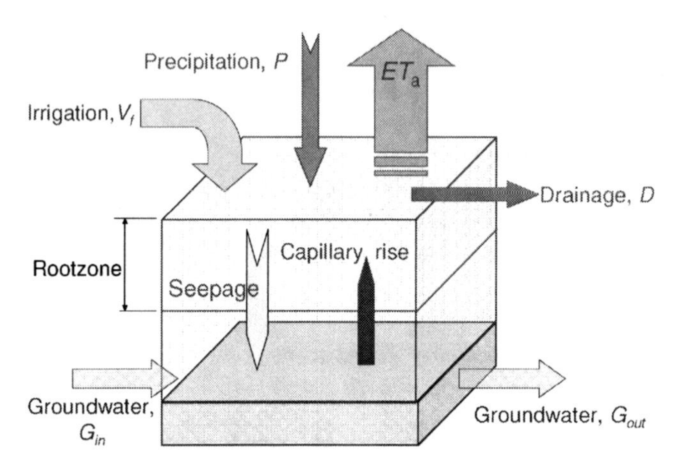

Fig. 1.7 Schematic water balance in an irrigated field (Bos 1984)

- Thirdly, the efficient use of irrigation water demands that the volume of diverted water is as practically low as possible. Precipitation on the area should be used as effectively as practical (Chapter 3).

As mentioned before, Chapter 5 shows how the limits on ET_a / ET_p in the fields, on the stability of the groundwater table under the command area, and on the use of water resource, result to the recommendation to manage the inflow of irrigation water in such a way that the depleted fraction of the gross area ranges between 0.5 and 0.9.

1.5 Calculating the Crop Irrigation Water Requirements

Chapter 6 describes a computer program that calculates the **cr**op **i**rrigation **wa**ter **r**equirements (CRIWAR) per user selected period (e.g. month, week, day, etc.) of a cropping pattern in an irrigated area, for various stages of crop development throughout the crops' growing season. The crop irrigation water requirements consist of the potential evapotranspiration, ET_p, minus the effective precipitation, P_e. Where, the potential evapotranspiration, ET_p was defined in Section 1.1. For the effective precipitation we use the definition that corresponds with the ICID terminology on the 'field application ratio' and the related efficiencies of water use at crop production level (Bos 1980; Bos and Nugteren 1974; ICID 1978):

Effective precipitation is that part of total precipitation on the cropped area, during a specific time period, which is available to meet evapotranspiration in the cropped area.

As will be explained in Chapters 2 and 6, CRIWAR calculates the ET_p on the basis of three (user-selected) alternative methods of estimating the reference evapotranspiration, ET_0. CRIWAR uses the equation:

$$ET_p = K_c \times ET_0 \qquad 1.4$$

As will be discussed in Section 2.9, the value of the crop coefficient, K_c, varies with the crop and the growth stage.

The method used to estimate ET_0 depends on the availability of accurate meteorological data and on local practices. As shown in Table 1.1, the FAO Modified Penman Method and the Penman-Monteith Method need the full range of accurate meteorological data (Penman 1948; Doorenbos and Pruitt 1977; Monteith 1965). Of these two methods, the FAO Modified Penman method is most widely used. However, as will be discussed in Chapter 2, the Penman-Monteith method is the recommended 'default' method (Burt et al. 2002; Allen et al. 1998).

If 'non-experts' collect meteorological data, the accuracy of the more advanced parameters (such as humidity, wind speed and radiation) can be very low or data can be missing. Under these conditions it is recommended to choose the Hargreaves-Samani Method rather than attempting to setup a complex data collection system or to 'repair' data series. As will be discussed in Chapter 2, this method gives a relatively good estimate of the reference *ET* (Hargreaves 1994; Droogers and Allen 2002).

As mentioned above, the calculated ET_p-value is reduced by the effective part of the precipitation, P_e to find the crop irrigation water requirements ($ET_p - P_e$). In order to calculate the effective precipitation, CRIWAR uses two semi-empirical methods. In addition CRIWAR allows the user to set the effective precipitation as a fixed percentage of total precipitation. The three methods are:

- The method as developed by the U.S. Department of Agriculture (1970). This method can be used if monthly precipitation data is available, as described in Section 3.3.
- A method based on the Curve Number Method as developed by the U.S. Soil Conservation Service (1964 and 1972). This method requires daily precipitation data, as described in Section 3.4.
- The user sets P_e as a percentage of P while P_e cannot exceed ET_p during the considered calculation period.

Table 1.1 Data requirements for methods to estimate the reference evapotranspiration (Droogers and Allen 2002)

Data needed	FAO Modified Penman	Penman-Monteith	Hargreaves-Samani
Minimum temperature	✓	✓	✓
Maximum temperature	✓	✓	✓
Humidity	✓	✓	
Wind speed	✓	✓	
Radiation	✓	✓	
Precipitation			✓

1.6 ET_p - P_e in the Field Water Balance

In estimating the crop irrigation water requirements ($ET_p - P_e$) we deal with two estimated parameters. Firstly, the ET_a usually is less than ET_p while it varies greatly from place to place. Figure 1.8 shows the spatial variation of ET_a in an irrigated area. As discussed in Section 1.3 and Chapter 5, a lower ET_a / ET_p ratio during dry periods may be part of the 'intended' scenario to reach a high productivity (in kg/m^3) of the applied irrigation water. Secondly, there are a variety of reasons why the effective precipitation is less than the actual precipitation (Chapter 3). Thus, in reality

$$ET_p - P_e > ET_a - P \tag{1.5}$$

As seen in Fig. 1.7, the ET_a and the precipitation, P, are the two components of the water balance that quantify the interface between soil-crop and atmosphere. In reality, both sides of the above equation are almost equal if the crop has no water stress ($ET_a \cong ET_p$) while there is no precipitation. In all other cases, the estimated crop water requirement will be greater than the net flux of water into the atmosphere. Chapter 5 deals with this issue by estimating the water required to maintain a stable groundwater table within a gross command area.

As shown in Fig. 1.7, one of the three inflows that balance ET_a is capillary rise. This capillary rise into the root zone depends heavily on site-specific soil physical conditions and on the rooting depth of the crop (which is a function of crop variety, soil type, groundwater depth, and climate). Such information is not normally known for any crop during its growth season. The capillary rise of groundwater

Fig. 1.8 Variation of ET_a (where ET_p = 6 mm/day) for an agricultural area of 650 ha, 31 May 2003, Hupselse Beek, The Netherlands

from the saturated zone into the root zone is estimated in the water management strategy part of CRIWAR.

If the depth to the groundwater table is shallow (less than 3 m) and the soil is fine-textured, capillary rise can contribute a significant volume of water to the root zone. However, for the groundwater table to remain stable, there must be a lateral flow of groundwater into the irrigated area; otherwise, capillary rise will decrease with the falling groundwater table. Although groundwater flow is not simulated in CRIWAR, the capillary component is corrected for in the crop irrigation water requirements via the water balance of the irrigated area (the depleted fraction as discussed in Chapter 4). Chapter 4 further explains how to correct the calculated crop water requirements for the contribution from groundwater.

1.7 The Irrigation Ratios

Crop irrigation water requirements ($ET_p - P_e$) should be transferred into irrigation water requirements at three levels of inlet structures that control irrigation water. Moving from the crop in an upstream direction, these inlets are:

- The field inlet controls the volume of water applied to a field. The ratio of ($ET_p - P_e$) over this volume is the 'field application ratio'. This ratio depends on the quality of the water application method and on the management skills of the irrigator (Section 5.2).
- The supply structure of the distribution system controls the volume of water that is supplied from the conveyance system (Fig. 1.9) to the distribution system. The distribution system is commonly managed by one large farmer (water user) or by a group of smaller farmers (association of water users). The degree by which this supplied volume needs to exceed the summed field application depends mainly on the dependability and uniformity of water distribution (Section 5.3).
- The head inlet structure controls the diverted flow from the surface water source into the conveyance system. The ratio between all supplied and the diverted water depends mainly on the quality of the system (seepage) and on the quality of management (misallocation of water). Both are influenced by the size of the irrigable area (Section 5.4).

1.8 Structure of the Handbook

The content of this book is organized in an 'upstream' direction. It starts with the evapo-transpiration of all crops (in the cropping pattern) within the considered irrigated area (Chapter 2). Chapter 3 discusses the factors that influence the fraction of the precipitation that can be consumed by the crop (effective precipitation). Chapter 4 covers the theory on capillary rise and demonstrates the potential contribution of

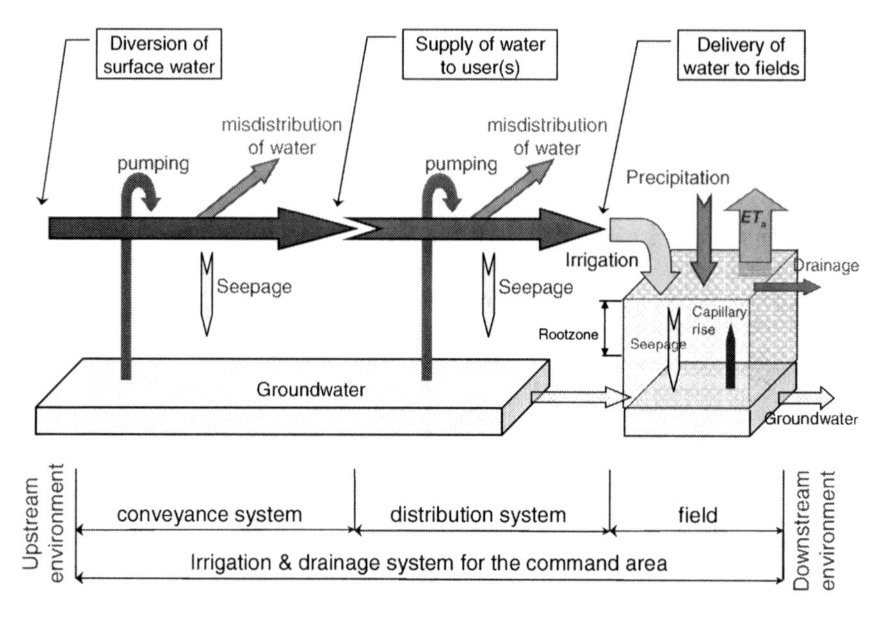

Fig. 1.9 Schematic flows of water within an irrigation and drainage system

capillary rise to soil moisture in the effective root zone. Chapter 5 presents a water management strategy for sustainable agriculture and explains the influence of the design and management of the irrigation (conveyance and distribution) system on the irrigation water requirements. The last chapter of the book contains the CRIWAR 3.0 user manual.

The book is thus comprised of the following six chapters:

1. Introduction
2. Evapotranspiration
3. Effective precipitation
4. Capillary rise
5. Irrigation water requirements
6. CRIWAR user's manual

Chapter 2
Evapotranspiration

2.1 Introduction

Evapotranspiration (ET) is an important term in the water balance of a cropped area. Irrigation engineers need to know how much of the applied irrigation water is consumed by the crop; only then can they calculate, or estimate, the remaining components of the water balance. Agriculturists, on the other hand, need to know the specific water requirements of a crop so that they can obtain a satisfactory yield; they also need to know whether these water requirements are being met under the prevailing irrigation practice. Reduction in ET due to plant-water stress, caused by water shortage, is associated with reduced plant yield, since both ET and photosynthesis are functions of stomatal regulation. Figure 2.1 shows the impact of the actual ET on above-ground biomass production (and thus on the yield) of a crop.

The method to estimate reference crop evapotranspiration (for cool season clipped grass) depends on the availability of accurate meteorological data and on local practices. As shown in Table 2.1, the FAO Modified Penman Method and the Penman-Monteith Method need a full range of accurate meteorological data (Penman 1948; Doorenbos and Pruitt 1977; Monteith 1965). Of these two methods the FAO Modified Penman method of Doorenbos and Pruitt 1977 has been widely used in the past. However, as will be discussed in Section 2.5, the Penman-Monteith method is the currently recommended "default" method (Burt et al. 2002; Allen et al. 1998). If, however, data on humidity, wind speed and radiation are missing the Hargreaves-Samani (1985) Method is recommended. This method gives a relatively good estimate of the reference *ET* (Section 2.4). How the theory on evapotranspiration is applied in practice is explained in Sections 2.6, 2.8, and 2.9.

The standardized Penman-Monteith (PM) ET_p equation of FAO 56 is the most commonly used reference ET method today and is appropriate for irrigation systems design and operation under a wide range of application situations and climates. The PM equation has been standardized to estimate the reference *ET* of both 12-cm tall, cool-season grass and 50 cm tall alfalfa (Allen et al. 2005a. The potential ET of a crop is estimated by multiplying the reference ET estimate by a crop factor specific to that crop and stage of growth. The $K_c \times ET_0$ method empirically incorporates many of the physiological and physical variables governing crop evapotranspiration.

M.G. Bos et al. *Water Requirements for Irrigation and the Environment*,

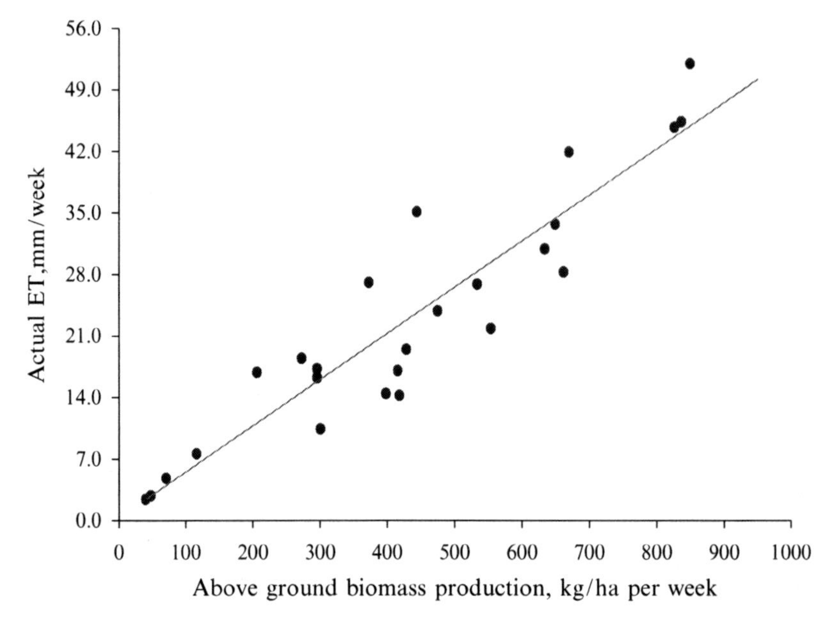

Fig. 2.1 Influence of ET_a on above-ground biomass production of cotton

Table 2.1 Data requirements for methods to estimate the reference evapotranspiration

Data needed	FAO Modified Penman	Penman-Monteith	Hargreaves-Samani
Minimum temperature	✓	✓	✓
Maximum temperature	✓	✓	✓
Humidity	✓	✓	
Wind speed	✓	✓	
Radiation	✓	✓	
Precipitation			✓

When applied carefully, the method produces estimates of ET_p that are sufficiently accurate for irrigation systems design and operation.

2.2 Developments in Theory

In the past, many empirical equations have been developed to estimate the potential crop evapotranspiration (i.e. the evapotranspiration from cropped soils that have an optimum water supply) (Blaney and Criddle 1950; Turc 1954; Jensen and Haise 1963a). These empirical correlation methods are often valid only for the local conditions under which they were developed, and as such are difficult to transfer to other areas. Nowadays, the focus is therefore on physically-based approaches, which have a wider applicability.

For the process of potential crop evapotranspiration, three basic physical requirements in the soil-plant-atmosphere system must be met:

- A continuously available supply of water
- Energy to change liquid water into vapour and
- A vapour gradient to maintain a flux from the evaporating surface to the atmosphere

The various methods of determining evapotranspiration are based on one or more of these requirements. For example, the soil-water-balance approach is based on 1, the energy-balance approach is based on 2, and the combination method (energy balance plus heat and mass transfer) is based on parts of 2 and 3. Penman (1948) was the first to introduce the combination method. He estimated the evaporation from an open water surface, and then used that as reference evaporation. Multiplied by a crop coefficient, this provided an estimate of the potential evapotranspiration from a cropped surface. Penman's Method requires meteorological data on air temperature, air humidity, solar radiation, and wind speed (Table 2.1). Because even this combination method contains a number of empirical relationships, a host of researchers have proposed numerous modifications to adjust it to local conditions.

After analyzing a range of lysimeter data world-wide, Doorenbos and Pruitt (1977) proposed the FAO Modified Penman Method, which has found world-wide application in irrigation and drainage projects. To estimate crop water requirements, CRIWAR uses the same two-step approach as Penman did, but it does not use Penman's open water evaporation, but the evapotranspiration from a reference crop. For the FAO Modified Penman Method (Doorenbos and Pruitt 1977), the reference crop was defined as:

An extended surface of an 0.08 to 0.15 m tall green grass cover of uniform height, actively growing, completely shading the ground, and not short of water.

There was evidence, however, that the Modified Penman Method over-predicted the crop water requirements (Jensen et al. 1990). Using similar physics as Penman did, Monteith (1965) developed an equation that describes the transpiration from a dry, extensive, horizontal, and uniformly vegetated surface, fully covering the ground that is optimally supplied with water. In international literature, this equation is known as the Penman-Monteith Equation.

Comparative studies (e.g. Jensen et al. 1990) showed the convincing performance of the Penman-Monteith approach under varying climatic conditions, thereby confirming the results of many individual studies reported over the past years. An expert consultation on procedures to revise the prediction of crop water requirements was held in Rome (Smith 1990). There, the consultation agreed to recommend the Penman-Monteith approach as the best-performing combination equation. Through the introduction of canopy and air resistances to water vapour diffusion, estimates of potential and actual evapotranspiration are, in principle, possible with the Penman-Monteith Equation.

Nowadays, this direct, or one-step, approach is increasingly being followed, especially in research environments. Nevertheless, since accepted canopy and air resistances may not yet be known for many crops, the two-step Penman approach (i.e. using crop factors multiplied by the reference crop ET) is still commonly used under field conditions. The grass reference crop in the Penman-Monteith approach is defined as (Allen et al. 1994):

> *A hypothetical crop fully covering the ground, and not short of water, with an assumed crop height of 0.12 m, a fixed canopy resistance of 70 s/m, and a canopy reflection coefficient of 0.23.*

This clipped, cool-season grass reference crop is widely accepted as the reference crop by both researchers and practitioners. Also the Hargreaves-Samani method uses this reference crop (Hargreaves 1994).

As mentioned in Section 2.1 evapotranspiration is an important factor in the water balance of an (irrigated) area. The above methods all estimate potential evapotranspiration. If sufficient water is available to meet the sum of evaporation and transpiration, the ET will reach its (maximum) potential value, ET_p, which is typically calculated as $K_c \times ET_0$. Otherwise, the actual evapotranspiration (ET_a) will be less than ET_p. As shown in Fig. 2.2, ET_a is commonly less than its potential value. This occurs if irrigation water or rainfall does not keep the root zone of the crop sufficiently hydrated during the growing season. Also evaporation (up to 2 mm/day) occurs before and after the cotton season. Thus, water also leaves the (irrigated) area if no crop is grown.

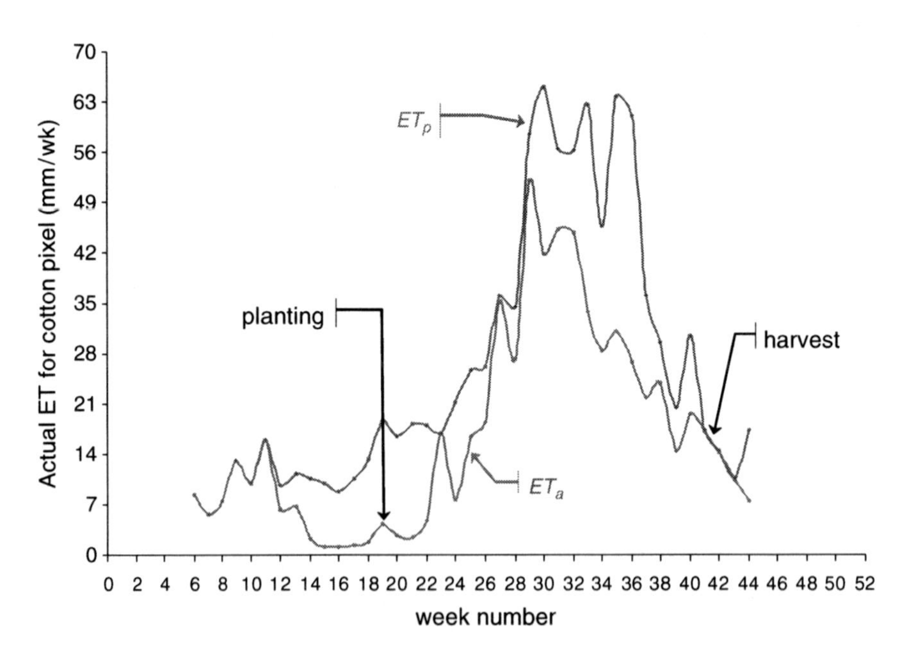

Fig. 2.2 Actual and potential evapotranspiration for irrigated cotton. ET_a is calculated from the energy balance of a satellite pixel, the ET_p using Penman-Monteith

2.2.1 *Evaporation from Open Water: The Penman Method*

As was mentioned earlier, the "classical" Penman Method (1948) and the FAO Modified Penman Method are generally no longer recommended, and the 'modern' and standardized Penman-Monteith Method (Section 2.3) is widely used around the globe as a standard (and default) method. Practitioners may proceed to Section 2.3. To give students a better understanding of the matter, however, we shall explain the original Penman Method.

Penman applied the energy balance of open water at the earth's surface. Equating all incoming and outgoing energy fluxes (Fig. 2.3), he obtained

$$R_n - G = H + \lambda E \qquad 2.1$$

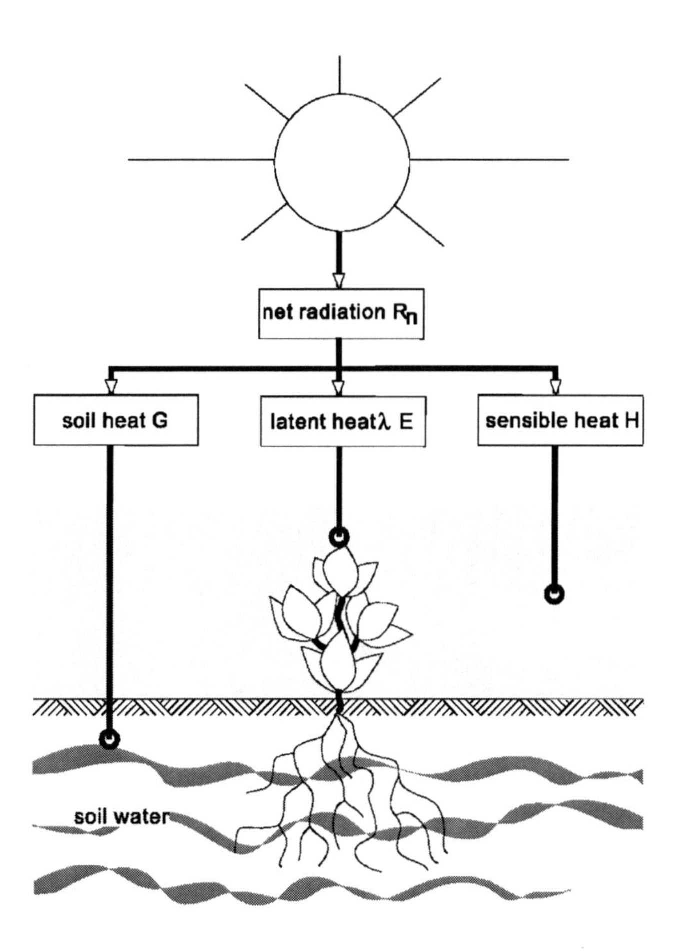

Fig. 2.3 Illustrating the variables involved in the energy balance of the soil surface (Feddes and Lenselink 1994)

where

R_n = energy flux density of net incoming radiation (W/m^2)
H = flux density of sensible heat into the air (W/m^2)
λE = flux density of latent heat into the air (W/m^2)
G = heat flux density into a water body or ground (W/m^2)

The coefficient λ in λE is the latent heat of vaporization of water and E is the vapour flux density (kg/m^{2}s). To convert λE (W/m^2) into an equivalent evapo(transpi)ration in units of mm/day, we multiply λE by a factor 0.0353. This factor equals the number of seconds in a day (86,400), divided by the value of λ (2.45 × 10^6 J/kg at 20°C), whereby we assume a density of water of 1,000 kg/m^3. Supposing that R_n and G can be measured, we can calculate E if we know the ratio $H/\lambda E$ (which is called the Bowen Ratio). We can derive this ratio from the transport equations of heat and water vapour in the air.

The situation shown in Fig. 2.2 and described by Equation 2.1 shows that radiation energy (R_n – G) is transformed into sensible heat, H, and water vapour, λE, which are transported to the air in accordance with

$$H = \rho_a c_p \frac{(T_s - T_z)}{r_a} \qquad 2.2$$

and, assuming a saturated surface:

$$\lambda E = \frac{\varepsilon \rho_a \lambda}{p_a} \times \frac{(e_{s,sat} - e_z)}{r_a} \qquad 2.3$$

where

c_p = specific heat of dry air at constant pressure (J/kgK)
ε = ratio of molecular masses of water vapour over dry air (dimensionless)
p_a = atmospheric pressure (kPa)
ρ_a = density of moist air (kg/m^3)
r_a = Aerodynamic diffusion resistance, assumed to be the same for heat and water vapour (s/m)

The other symbols are illustrated in Fig. 2.4:

T_s = temperature at the evaporating (water or leaf/soil) surface (°C)
T_z = air temperature at a height z above the surface (°C)
$e_{s,sat}$ = saturation vapour pressure at the evaporating (water) surface (kPa)
e_z = prevailing vapour pressure in the external air, measured at the same height as T_z (k/Pa)

Applying the concept of the similarity of the transport of heat and of water vapour yields the Bowen Ratio, β:

$$\beta = \frac{H}{\lambda E} = \frac{c_p p_a}{\lambda \varepsilon} \times \frac{T_s - T_z}{e_{s,sat} - e_z} \qquad 2.4$$

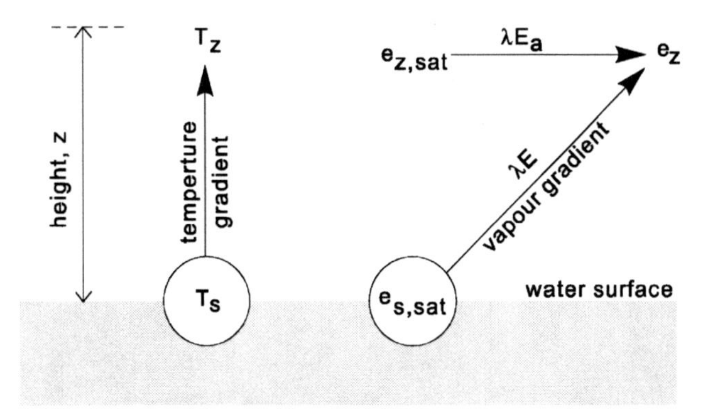

Fig. 2.4 Illustration of terminology

where the ratio $c_p p_a/\lambda\varepsilon$ is commonly replaced by γ, termed the psychrometric constant (kPa/°C). The problem with the above equations is that the surface temperature, T_s, is not generally known (not routinely measured). Penman therefore took three intermediate steps:

1. He introduced the proportionality constant

$$\Delta = \frac{e_{s,sat} - e_{z,sat}}{T_s - T_z} \tag{2.5}$$

The proportionality constant Δ (kPa/°C) is the first derivative of the function $e_{z,sat}$ versus T_z, known as the saturation vapour pressure curve (Fig. 2.5). Note that $e_{s,sat}$ in Equation 2.5 is the saturation vapour pressure at the surface at temperature T_z. Hence

$$\Delta = \frac{de_z}{dT_z} = \frac{e_{s,sat} - e_{z,sat}}{T_s - T_z} \tag{2.6}$$

Substituting Equation 2.5 into Equation 2.4 yields

$$\beta = \frac{H}{\lambda E} = \frac{\gamma}{\Delta} \times \frac{e_{s,sat} - e_{z,sat}}{e_{s,sat} - e_z} \tag{2.7}$$

2. He replaced the vapour pressure gradient $e_{s,sat} - e_{z,sat}$ in Equation 2.7 with

$$(e_{s,sat} - e_z) - (e_{z,sat} - e_z)$$

This gives

$$\beta = \frac{\gamma}{\Delta}\left(1 - \frac{e_{z,sat} - e_z}{e_{s,sat} - e_z}\right) \tag{2.8}$$

3. He defined an 'adiabatic vapour transport' term that occurs if $e_{s,sat} \approx e_{z,sat}$. If we introduce this assumption into Equation 2.3, the theoretical adiabatic evaporation, λE_a, equals

$$\lambda E_a = \frac{\varepsilon \rho_a \lambda}{p_a} \times \frac{e_{z,sat} - e_z}{r_a} \tag{2.9}$$

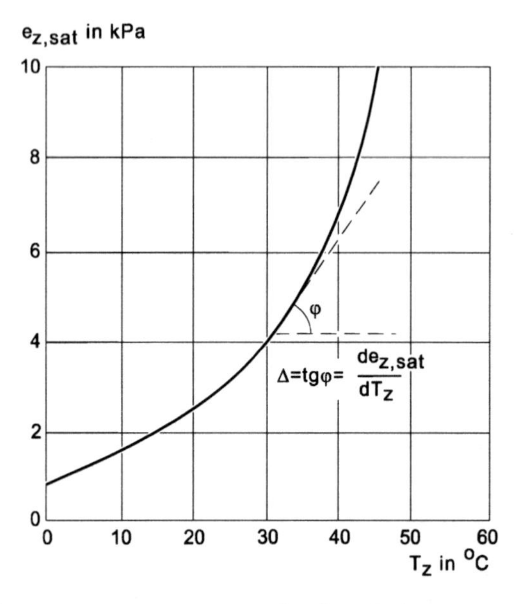

Fig. 2.5 Saturated vapour pressure, $e_{z,sat}$, as a function of air temperature, T_z (Feddes and Lenselink 1994)

A comparison of this equation with Equation 2.3 shows that

$$\frac{e_{z,sat} - e_z}{e_{s,sat} - e_z} = \frac{E_a}{E} \qquad 2.10$$

so that

$$\beta = \frac{\gamma}{\Delta}\left(1 - \frac{E_a}{E}\right) \qquad 2.11$$

Substituting the above information into Equation 2.1, and writing E_o (subscript o denoting open water) for E yields the Penman Formula, which is

$$E_o = \frac{\Delta}{\Delta + \gamma} \times \frac{R_n - G}{\lambda} + \frac{\gamma}{\Delta + \gamma} E_a \qquad 2.12$$

where, as defined above:

E_o = open water evaporation rate (kg/m²s)
Δ = proportionality constant de_z/dT_z (kPa/°C)
R_n = net radiation (W/m²)
G = heat flux density into the water body (W/m²)
λ = latent heat of vaporization (J/kg)

γ = psychrometric constant (kPa/°C)
E_a = isothermal evaporation rate (kg/m^2 s)

Equation 2.12 shows the combination of two processes in one equation. The first term is some fraction of the evaporation equivalent of the net flux of radiant energy to the surface, also called the 'radiation term'. The second term quantifies the corresponding aerodynamic process of water-vapour transport from the evaporating water surface to the surrounding air, also called the 'aerodynamic term'. Note that the resulting E_0 (kg/m^2 s) should be multiplied by 86,400 seconds to give the equivalent evaporation rate E_0 in mm/day.

The Penman-Monteith method, which is similar to the Penman method, except with the addition of a surface resistance term and the replacement of the empirical Penman wind function with a theoretically aerodynamic resistance term, can be derived in a similar manner as above.

As was mentioned in Section 2.2, the original Penman Formula (Equation 2.12) used E_0 as reference evaporation. The practical value of estimating E_0 with Equation 2.12, however, is generally limited to large water bodies (e.g. lakes and flooded rice fields in the very early stages of growth), and the G term for open water is difficult to estimate. Following its development, the Penman equation was applied to grassed surfaces for ET_0 (Penman 1984). But, as was also mentioned earlier, we do not use Equation 2.12.

2.2.2 *The FAO Modified Penman Method*

2.2.2.1 The Modification

The modification of the Penman Method, as introduced by Doorenbos and Pruitt (1977), started from the assumption that evapotranspiration from grass largely occurs in response to climatic conditions. Because short grass is the common surface cover surrounding agro-meteorological stations, they suggested that, instead of using evaporation from open water as a reference, the evapotranspiration from grass, 0.08–0.15 m tall and not short of water, be used. The main changes in Penman's Formula to compute this reference evapotranspiration relate to:

- The short-wave reflection coefficient (approximately 0.05 for water and 0.25 for grass).
- A more sensitive wind function in the aerodynamic term.
- An adjustment factor to take into account that local climatic conditions deviate from an assumed standard. This adjustment is needed to allow various combinations of radiation, relative humidity, and day/night wind ratios (see Table 2.2).

If the heat flux, G, is set equal to zero for daily periods, which is generally valid for a fully cover vegetated surface, the Modified Penman Equation can be written as

$$ET_g = c\left[\frac{\Delta}{\Delta+\gamma} \times 86400\,\frac{R_n}{\lambda} + \frac{\gamma}{\Delta+\gamma}\,2.7\,f(u)(e_{z,sat} - e_z)\right] \qquad 2.13$$

where

ET_g = reference evapotranspiration rate (mm/day)
c = dimensionless adjustment factor (see Section 2.4.2)
R_n = energy flux density of net incoming radiation (W/m^2)
$f(u)$ = wind function; $f(u) = 1 + 0.864u_2$
u_2 = wind speed at 2.0 m above ground surface (m/s)
$e_{z,sat} - e_a$ = vapour pressure deficit (kPa)
Δ, γ = as defined earlier

Potential evapotranspiration from a cropped surface is subsequently found by multiplying this reference, ET_g, by the appropriate crop coefficient (Section 2.6).

2.2.2.2 The Adjustment Factor, c

If the average climatological conditions for which the (Modified) Penman Formula was developed are not met, the adjustment factor in Equation 2.13 differs

Table 2.2 Adjustment factor, c, as a function of the maximum relative humidity, RH_{max}, incoming shortwave radiation, R_s, day-time wind speed, u_{day}, and the wind speed ratio, u_{day}/u_{night} (Doorenbos and Pruitt 1977)

R_s	RH_{max} = 30%				RH_{max} = 60%				RH_{max} = 90%			
(mm/day)	3	6	9	12	3	6	9	12	3	6	9	12
u_{day} (m/s)	u_{day}/u_{night} = 4.0											
0	.86	.90	1.00	1.00	.96	.98	1.05	1.05	1.02	1.06	1.10	1.10
3	.79	.84	.92	.97	.92	1.00	1.11	1.19	.99	1.10	1.27	1.32
6	.68	.77	.87	.93	.85	.96	1.11	1.19	.94	1.10	1.26	1.33
9	.55	.65	.78	.90	.76	.88	1.02	1.14	.88	1.01	1.16	1.27
	u_{day}/u_{night} = 3.0											
0	.86	.90	1.00	1.00	.96	.98	1.05	1.05	1.02	1.06	1.10	1.10
3	.76	.81	.88	.94	.87	.96	1.06	1.12	.94	1.04	1.18	1.28
6	.61	.68	.81	.88	.77	.88	1.02	1.10	.86	1.01	1.15	1.22
9	.46	.56	.72	.82	.67	.79	.88	1.05	.78	.92	1.06	1.18
	u_{day}/u_{night} = 2.0											
0	.86	.90	1.00	1.00	.96	.98	1.05	1.05	1.02	1.06	1.10	1.10
3	.69	.76	.85	.92	.83	.91	.99	1.05	.89	.98	1.10	1.14
6	.53	.61	.74	.84	.70	.80	.94	1.02	.79	.92	1.05	1.12
9	.37	.48	.65	.76	.59	.70	.84	.95	.71	.81	.96	1.06
	u_{day}/u_{night} = 1.0											
0	.86	.90	1.00	1.00	.96	.98	1.05	1.05	1.02	1.06	1.10	1.10
3	.64	.71	.82	.89	.78	.86	.94	.99	.85	.92	1.01	1.05
6	.43	.53	.68	.79	.62	.70	.84	.93	.72	.82	.95	1.00
9	.27	.41	.59	.70	.50	.60	.75	.87	.62	.72	.87	.96

from 1.0. The values of the adjustment factor, c, can be estimated from comparisons of calculated and measured values of ET_g, whereby the interactions between wind speed, relative humidity, and solar radiation are analyzed. Table 2.2 gives values of c as a function of the day-time wind speed, u_{day}, the ratio of day over night wind speed, (u_{day}/u_{night}), the maximum relative humidity, RH_{max}, and the solar radiation, R_s.

Data need to be supplied on the day-night ratio of the wind speed and on the maximum relative humidity. However, if these data are not available, CRIWAR will use the following default values:

$$u_{day} / u_{night} = 2.0$$

and

$$RH_{max} = (RH + 100) / 2$$

where *RH* is the average relative humidity. Day-time wind speed is calculated from data on mean wind speed and a day-night wind ratio. The calculation of incoming short-wave radiation is explained in Section 2.6.

2.3 The Penman-Monteith Approach

2.3.1 The Equation

In analogy with Section 2.2, the evapotranspiration from a cropped surface having full water supply can be described by an equation very similar to Equation 2.13. Nevertheless, we have to take into account the differences between the grassed surface of Doorenbos and Pruitt and the hypothetical reference crop surface of FAO-56. In this context, these differences are:

- The albedo (or reflection coefficient for solar radiation) is different for the hypothetical reference crop surface (0.23) and the grassed surface of Doorenbos and Pruitt (0.25).
- The hypothetical reference crop surface has a roughness (dependent on crop height and wind speed). The method uses a theoretical aerodynamic function and an explicit roughness term rather than using the empirical wind function of Doorenbos and Pruitt that was fitted to grass.
- A stomatal diffusion resistance is added, resulting to a modification of the psychrometric constant, γ.

As mentioned before, grass reference ET_0 was defined in FAO-24 (Doorenbos and Pruitt 1977) as "the rate of evapotranspiration from an extensive surface of 8–15 cm tall, green grass cover of uniform height, actively growing, completely shading the ground and not short of water." It is generally accepted that the grass reference crop is a 'cool-season', C-3 photosynthetic-pathway grass (where the CO_2 is first incorporated into a 3-carbon compound) with roughness, density, leaf area and bulk surface resistance characteristics similar to perennial ryegrass (*Lolium perenne*) or (clipped) tall fescue (*Festuca arundinacea* Schreb.). Because of the challenges in growing and maintaining a living reference crop, the PM equation with defined canopy and aerodynamic resistances was adopted by the United Nations – Food and Agriculture Organization (FAO) (Smith et al. 1991, 1996; Allen et al. 1998, 2005a) and the ASCE-EWRI (2005) to represent a standardized ET_0. The FAO definition for ET_0 in terms of the PM equation is "the rate of evapotranspiration from a hypothetical reference crop with an assumed crop height of 0.12 m, a fixed surface resistance of 70 s m^{-1} and an albedo of 0.23, closely resembling the evapotranspiration from an extensive surface of green grass of uniform height, actively growing, completely shading the ground and with adequate water" (Allen et al. 1998). The ASCE-EWRI (2005) adopted the same definition for standardization of ET_0, with the provision for lower surface daytime resistance (50 s m^{-1}) and a higher surface nighttime resistance (200 s m^{-1}), when calculating on hourly or shorter time steps. The ASCE-EWRI resistances for hourly calculations were subsequently adopted by FAO (Allen et al. 2005a).

The grass reference ET_0 is utilized in this chapter as the basis for the $K_c \times ET_0$ method due to its traditional use in water requirement estimates for landscapes. In addition, ET_0 is used as the basis for water requirement estimates for agricultural crops in Europe, Africa, Asia, and in a number of US states, particularly in coastal, southern and eastern parts of the US. Alfalfa reference ET_r is widely used as an ET reference basis in a number of western US states, and in some instances represents a superior reference due to its height and high leaf area (Pereira et al. 1999). However, the alfalfa reference is not introduced in this chapter for reasons of brevity. The alfalfa reference is described in detail in Jensen et al. (1990), ASCE-EWRI (2005), Jensen et al. (2007) and Allen et al. (2007).

The ASCE-EWRI (2005) standardized PM method for grass reference ET_0 has a condensed, simplified form from the original PM method:

$$ET_0 = \frac{0.408\,\Delta(R_n - G) + \gamma \dfrac{C_n}{T+273} u_2\,(e_s - e_a)}{\Delta + \gamma\,(1 + C_d\,u_2)} \qquad 2.14$$

Where

ET_0 = standardized reference ET for a 12 cm tall, cool season grass in mm day^{-1} for daily time steps or mm h^{-1} for hourly time steps

R_n = calculated net radiation at the crop surface in MJ m^{-2} day^{-1} for daily time steps or MJ m^{-2} h^{-1} for hourly time steps

G = soil heat flux density at the soil surface in MJ m^{-2} day^{-1} for daily time steps or MJ m^{-2} h^{-1} for hourly time steps

T = mean daily or hourly air temperature at 1.5–2.5 m height (°C)
u_2 = mean daily or hourly wind speed at 2-m height (ms^{-1})
e_s = saturation vapor pressure at 1.5–2.5 m height (kPa), calculated for daily time steps as the average of saturation vapor pressure at maximum and minimum air temperature
e_a = mean actual vapor pressure at 1.5–2.5m height (kPa) Δ slope of the saturation vapor pressure versus temperature curve (kPa $°C^{-1}$)
Δ = slope of the saturation vapor pressure versus temperature curve (kPa $°C^{-1}$)
γ = psychrometric constant (kPa $°C^{-1}$)
C_n = numerator constant that changes with reference type and calculation time step (K mm s^3 Mg^{-1} d^{-1} or K mm s^3 Mg^{-1} h^{-1})
C_d = denominator constant that changes with reference type and calculation time step (s m^{-1})

Units for the 0.408 coefficient are m^2 mm MJ^{-1} [this coefficient embodies the latent of vaporization, λ, and water density, ρ_w; where $\lambda = 2.45$ MJ kg^{-1} and $\rho_w = 1.0$ Mg m^{-3}].

Table 2.3 provides values for C_n and C_d for standardized ET_0. The values for C_n consider the time step and aerodynamic roughness of the 12 cm grass surface. The constant in the denominator, C_d, considers the time step, bulk surface resistance, and aerodynamic roughness of the surface, time step and daytime/nighttime). C_n and C_d were derived by simplifying several terms within the ASCE-PM equation of ASCE Manual 70 (Allen et al. 1989; Jensen et al. 1990) and rounding the result. Daytime is defined as occurring when R_n during an hourly period is positive. The ASCE-EWRI (2005) and the FAO (Allen et al. 2005a) definition use a smaller value for surface resistance for hourly or shorter calculation time steps (during daytime) than for daily calculation time steps. The daily FAO-PM (Allen et al. 1998) is equivalent to Equation 2.14, where $C_n = 900$ and $C_d = 0.34$. ET_0 estimates from Equation 2.14 are similar to those from the California Irrigation Management Information System (CIMIS) modified Penman ET_0 (ASCE-EWRI 2005; Allen et al. 2005a; Ventura et al. 1999).

In this chapter, only ET_0 is used as ET_{ref}, and the K_c values are based on ET_0. The alfalfa reference (ET_r) and appropriate K_c values are presented in Wright (1982), Jensen et al. (1990, 2007) and Allen et al. (2007).

Table 2.3 Values for C_n and C_d in Equation 2.14 (From ASCE-EWRI 2005; Allen et al. 2005a)

Calculation time step	Short Reference ET_0 (clipped grass)		Units for ET_0	Units for R_n, G	G/R_n Ratio
	C_n	C_d			
Daily	900	0.34	mm day^{-1}	MJ m^{-2} day^{-1}	0
Hourly during daytime	37	0.24	mm h^{-1}	MJ m^{-2} h^{-1}	0.1
Hourly during nighttime	37	0.96	mm h^{-1}	MJ m^{-2} h^{-1}	0.5

2.3.2 *Effect of Time Step Size on Calculations*

The Penman and Penman-Monteith equations can be applied to hourly and 24-h time steps. The 24-h time steps can use daily, weekly, 10-day, and monthly means for weather data. Under many climatic conditions, calculating ET_0 using hourly time steps and then summing over 24-h provides better estimates of ET_0 than as calculated using 24-h average data with 24-h calculation time steps (Itenfisu et al. 2003; ASCE-EWRI 2005; Allen et al. 2005a). Generally, calculating 24-h ET_0 by summing hourly or shorter time steps will improve the estimation accuracy because this captures the inherent co-variances in hourly wind speed, net radiation, and vapor pressure deficits that invariably occur at many sites (Irmak et al. 2005; Allen et al. 2005a). Examples of this are high wind conditions during afternoon with low humidity, overpass of cloud fronts and rain events, and nighttime calm or coastal sea breezes (common at late afternoon and early evening) at sites near large water bodies.

2.3.3 *Computation of Parameters for the Penman-Monteith Reference Equation*

It is recommended that standardized procedures and equations be used to calculate parameters in ET_{ref} equations. This insures agreement among independent calculations and simplifies calculation verification. Equations presented in this section follow procedures standardized by FAO-56 (Allen et al. 1998) and by the ASCE-EWRI (2005).

2.3.3.1 Saturation Vapor Pressure of the Air

For 24-h or longer calculation time steps, e_s, the saturation vapor pressure of the air, is computed as:

$$e_s = \frac{e^{o}(T_{max}) + e^{o}(T_{min})}{2} \tag{2.15}$$

where T_{max} and T_{min} are daily maximum and minimum air temperature, °C, at the measurement height (1.5–2 m), and e^o is the saturation vapor pressure function. For hourly applications, e_s is calculated as e^o(T) where T is average hourly air temperature.

$$e^{o}(T) = 0.6108\, exp\left(\frac{17.27\,T}{T + 237.3}\right) \tag{2.16}$$

where $e^o(T)$ is in kPa and T is in °C (Tetens 1930).

2.3.3.2 Actual Vapor Pressure of the Air

Actual vapor pressure of the air, e_a, is equivalent to saturation vapor pressure at the dew point temperature, T_d. For 24-h or longer time steps, T_d is taken as mean daily or early morning dew point temperature (°C). Humidity of the air can be measured using several methods, including relative humidity sensors, dew point sensors, and wet bulb/dry bulb psychrometers, so that e_a can be calculated many different ways. The recommended procedures, in order of what are considered to be the most reliable to the least reliable, are (ASCE-EWRI 2005):

1. For 24-h periods, averaging e_a measured or computed hourly over the 24-h period.
2. For 24-h periods, calculating e_a from dew point, T_d, that is measured or computed hourly over the 24-h period:

$$e_a = e^o(T_d) = 0.6108 \exp\left(\frac{17.27\, T_d}{T_d + 237.3}\right) \qquad 2.17$$

   where e_a is in kPa and T_d is in °C.

3. For hourly calculations, e_a is commonly calculated from RH as:

$$e_a = \frac{RH}{100} e^o(T) \qquad 2.18$$

   where RH is mean relative humidity for the hourly or shorter period, %, and T is mean air temperature for the hourly or shorter period, °C.
4. Psychrometer measurements using dry and wet bulb thermometers. Psychrometric procedures are described in Jensen et al. (1990), FAO-56 (Allen et al. 1998), and ASCE-EWRI (2005).
5. For 24-h or longer time steps, relative humidity (RH) measurements taken twice daily (early morning, corresponding to T_{min} and early afternoon, corresponding to T_{max}) can be combined to yield an approximation for 24-h average e_a:

$$e_a = \frac{e^o\left(T_{\min}\right)\dfrac{RH_{\max}}{100} + e^o\left(T_{\max}\right)\dfrac{RH_{\min}}{100}}{2} \qquad 2.19$$

   where RH_{max} is daily maximum relative humidity (%) (during early morning), and RH_{min} is daily minimum relative humidity (%) (during early afternoon, around 1400 h).
6. From daily RH_{max} and T_{min}:

$$e_a = e^o\left(T_{\min}\right)\frac{RH_{\max}}{100} \qquad 2.20$$

7. From daily RH_{min} and T_{max}:

$$e_a = e^o\left(T_{\max}\right)\frac{RH_{\min}}{100} \qquad 2.21$$

8. If daily humidity data are missing or are of questionable quality, e_a can be approximated for the reference environment assuming that T_d is near T_{min}:

$$T_d = T_{\min} - K_o \qquad 2.22$$

   where K_o is approximately 2–4°C in dry seasons in semiarid and arid climates K_o is approximately 0°C in humid to sub-humid climates (ASCE-EWRI 2005) and the rainy season in semi-arid climates.
9. In the absence of RH_{max} and RH_{min} data, but where daily mean RH data are available, e_a can be estimated as:

$$e_a = \frac{RH_{mean}}{100} e^o\left(T_{mean}\right) \qquad 2.23$$

   where RH_{mean} is mean daily relative humidity, generally defined as the average between RH_{max} and RH_{min}, and T_{mean} is mean daily air temperature. Equation 2.23 is less desirable than previous methods for e_a due to the non-linear e^o (T) versus T relationship.

2.3.3.3 Psychrometric Constant

The psychrometric constant (γ) in the Penman and PM equations is calculated following (Brunt 1952):

$$\gamma = 0.000665\ P \qquad 2.24$$

where P has units of kPa and γ has units of kPa °C^{-1}.

2.3.3.4 Atmospheric Pressure

For purposes of ET estimation, mean atmospheric pressure, P, is calculated from elevation (ASCE-EWRI 2005):

$$\mathrm{P} = \left(2.406 - 0.0000534\,\mathrm{z}\right)^{5.26} \qquad 2.25$$

where P is has units of kPa, and z is the weather station elevation above mean sea level in meters.

2.3.3.5 Slope of the Saturation Vapor Pressure-Temperature Curve

The slope of the saturation vapor pressure-temperature curve, Δ, is computed as:

$$\Delta = \frac{2503\ \exp\left(\frac{17.27\,T}{T+237.3}\right)}{(T+237.3)^2} \tag{2.26}$$

where Δ has units of kPa °C-1 and T is daily or hourly mean air temperature in °C.

2.3.3.6 Wind Speed at 2 m

Wind speed varies with height above the ground surface. For the calculation of standardized ET_{ref}, the wind speed measurement is considered to be 2 m above the grass surface. Therefore, one can adjust the wind speed measured at other heights using:

$$u_2 = u_z \frac{4.87}{\ln(67.8 z_w - 5.42)} \tag{2.27}$$

where u_2 is wind speed at 2 m above ground surface in m s^{-1}, u_z is measured wind speed at z_w m above ground surface in m s^{-1}, and z_w is the height of wind measurement above the ground surface in m. Equation 2.27 is used for measurements taken above a short grass (or similar) surface, based on the logarithmic wind speed profile equation. For wind speed measurements made above surfaces other than clipped grass, the user should apply a full logarithmic equation that considers the influence of vegetation height and roughness on the shape of the wind profile. These alternative adjustments are described in Allen and Wright (1997) and ASCE-EWRI (2005). Wind speed data measured at heights above 2 m are acceptable to use in the standardized equations following adjustment to the standard 2 m height by Equation 2.27, and wind speeds measured at taller heights are preferred if vegetation adjacent to the weather station commonly exceeds 0.5 m. Measurement at height above 2 m, reduces the influence of the taller vegetation surrounding the weather measurement site.

2.3.3.7 Net Radiation

Net radiation, R_n, in the context of ET, is the net amount of radiant energy available at a vegetation or soil surface for evaporating water, heating the air, or heating the surface. R_n is estimated from the short and long wave band components:

$$R_n = R_{ns} - R_{nl} \tag{2.28}$$

where R_{ns} is net short-wave radiation, [MJ m^{-2} day^{-1} or MJ m^{-2} h^{-1}], defined as being positive downwards and negative upwards, and R_{nl} is net outgoing long-wave radiation, [MJ m^{-2} day^{-1} or MJ m^{-2} h^{-1}], defined as being positive upwards and negative downwards. This sign convention produces values for R_{ns} and R_{nl} that are generally positive or zero.

Net short-wave radiation resulting from the balance between incoming and reflected solar radiation is given by:

$$R_{ns} = (1 - \alpha)\, R_s \qquad 2.29$$

where α is albedo, fixed in the FAO-56 (Allen et al. 1998) and ASCE-EWRI (2005) standardizations at 0.23 for both daily and hourly time steps [dimensionless] and R_s is incoming solar radiation [MJ m^{-2} day^{-1} or MJ m^{-2} h^{-1}]. The standardized ASCE-EWRI procedure for estimating R_{nl} is the same as that adopted by FAO-56 (Allen et al. 1998) and is based on the Brunt (1932, 1952) approach for estimating net emissivity and for daily time intervals:

$$R_{nl} = \sigma f_{cd}\left(0.34 - 0.14\sqrt{e_a}\right)\frac{T^4_{k\max} + T^4_{k\min}}{2} \qquad 2.30$$

For hourly time intervals:

$$R_{nl} = \sigma f_{cd}\left(0.34 - 0.14\sqrt{e_a}\right)T^4_{Khr} \qquad 2.31$$

where R_{nl} has units of [MJ m^{-2} day^{-1} or MJ m^{-2} h^{-1}], σ is the Stefan-Boltzmann constant [4.901×10^{-9} MJ K^{-4} m^{-2} day^{-1} and 2.042×10^{-10} MJ K^{-4} m^{-2} h^{-1}], f_{cd} is a cloudiness function [dimensionless] and limited to $0.05 \le f_{cd} \le 1.0$, e_a is actual vapor pressure [kPa] (see Equations 2.19–2.21, or 2.23), $T_{K\,max}$ is maximum absolute temperature during the 24-h period [K] (K = °C + 273.15), $T_{K\,min}$ is minimum absolute temperature during the 24-h period [K], and $T_{K\,hr}$ is mean absolute temperature during the hourly period [K]. The superscripts "4" in Equations 2.30 and 2.31 indicate the need to raise the absolute air temperature (K ≈ 273 + °C) to the power of 4.

For daily and monthly calculation time steps, f_{cd} is calculated as:

$$f_{cd} = 1.35\frac{R_s}{R_{so}} - 0.35 \qquad 2.32$$

where R_s/R_{so} is relative solar radiation, R_s is measured or calculated solar radiation [MJ m^{-2} day^{-1}], and R_{so} is calculated clear-sky radiation [MJ m^{-2} day^{-1}]. The ratio R_s/R_{so} is used to represent cloudiness and is limited to $0.3 < R_s/R_{so} \le 1.0$ so that f_{cd} has limits of $0.05 \le f_{cd} \le 1.0$.

For hourly periods during daytime when the sun is more than about 15° above the horizon, f_{cd} is calculated using Equation 2.32 with the same limits applied. For hourly periods during nighttime, R_{so}, by definition, equals 0, and Equation 2.32 is undefined. Therefore, f_{cd} during periods of low sun angle and during nighttime is defined using f_{cd} from prior periods just before sunset. When the sun angle[1] (β) above the horizon at the midpoint of the hourly or shorter time period is less than 0.3 radians (~17°), then (ASCE-EWRI 2005):

$$f_{cd} = f_{cd\beta>0.3} \quad 2.33$$

where $f_{cd\,\beta > 0.3}$ is the cloudiness function for the time period prior to when β falls below 0.3 radians during afternoon or evening [dimensionless].

If the calculation time step is shorter than 1 h, then f_{cd} from several periods can be averaged into $f_{cd\,\beta > 0.3}$ to obtain a representative average value. In mountain valleys where the sun may set near or above 0.3 radians (~17°), the user should increase the sun angle at which $f_{cd\,\beta > 0.3}$ is computed and imposed. For example, for a location where mountain peaks are 20° above the horizon, a period should be selected for computing $f_{cd\,\beta > 0.3}$ where the sun angle at the end of the time period is 25–30° above the horizon. The same adjustment is necessary when deciding when to resume computation of f_{cd} during morning hours when mountains lie to the east.

Only one value for $f_{cd\,\beta > 0.3}$ is calculated per day for use during dusk, nighttime and dawn periods. That value for $f_{cd\,\beta > 0.3}$ is then applied to the time period when β at the midpoint of the period first falls below 0.3 radians (~17°) and to all subsequent periods until after sunrise when β again exceeds 0.3 radians.

Equations 2.21 and 2.22 will not apply at latitudes and times of the year when there are no hourly (or shorter) periods having sun angle of 0.3 radians or greater. These situations occur at latitudes of 50° for about 1 month per year (in winter), at latitudes of 60° for about 5 months per year, and at latitudes of 70° for about 7 months per year (ASCE-EWRI 2005). Under these conditions, the application can average $f_{cd\,\beta > 0.3}$ from fewer time periods or, in the absence of any daylight, can assume a ratio of R_s/R_{so} ranging from 0.3 for complete cloud cover to 1.0 for no cloud cover. Under these extreme conditions, the estimation of R_n is only approximate.

2.3.3.8 Clear-Sky Solar Radiation (R_{so})

Clear-sky solar radiation (R_{so}) is used in the calculation of net radiation (R_n). R_{so} is defined as the amount of short-wave radiation that would be received at the weather measurement site under conditions of clear-sky (i.e., cloud-free). Daily R_{so} is a function of the time of year and latitude and is impacted by station elevation (affecting atmospheric thickness and transmissivity), the amount of precipitable water in the atmosphere (affecting the absorption of some short-wave radiation), and the amount of dust or aerosols in the air. R_{so}, for purposes of calculating R_n, can be computed as:

$$R_{so} = \left(0.75 + 2x10^{-5} z\right) R_a \quad 2.34$$

where z (m) is station elevation above sea level.

---

[1] The sun angle β is defined as the angle of a line from the measurement site to the center of the sun's disk relative to a line from the measurement site to directly below the sun and tangent to the earth's surface. This definition assumes a flat surface.

2.3.3.9 Exoatmospheric Radiation

Exoatmospheric radiation, R_a, also known as extraterrestrial radiation, is defined as the short-wave solar radiation in the absence of an atmosphere (or at the outer limits of the earth's atmosphere) and is used to calculate R_{so}. For daily (24-h) periods, R_a is estimated from the solar constant, the solar declination, and the day of the year:

$$R_a = \frac{24}{\pi} G_{sc} d_r [\omega_s \sin(\varphi)\sin(\delta) + \cos(\varphi)\cos(\delta)\sin(\omega_s)] \qquad 2.35$$

where R_a has units of [MJ m^{-2} day^{-1}], G_{sc} is the solar constant [4.92 MJ m^{-2} h^{-1}], d_r is the inverse relative distance factor (squared) for the earth-sun [dimensionless], ω_s is the sunset hour angle [radians], $\varphi = L\ (\pi/180)$ [radians] for latitude L in degrees, and δ is solar declination [radians]. The latitude, φ, is positive for the northern hemisphere and negative for the southern hemisphere.

Parameters d_r and δ are calculated as:

$$d_r = 1 + 0.033 \cos\left(\frac{2\pi}{365} J\right) \qquad 2.36$$

$$\delta = 0.409 \sin\left(\frac{2\pi}{365} J - 1.39\right) \qquad 2.37$$

where J is the number of the day in the year between 1 (1 January) and 365 or 366 (31 December). The constant 365 in Equations 2.36 and 2.37 is held at 365 even during a leap year. J can be calculated as:

$$J = D_M - 32 + Int\left(275\frac{M}{9}\right) + 2\ Int\left(\frac{3}{M+1}\right) + Int\left(\frac{M}{100} - \frac{Mod(Y,4)}{4} + 0.975\right) \qquad 2.38a$$

where D_M is the day of the month (1–31), M is the number of the month (1–12), and Y is the number of the year (for example 1996 or 96). The "Int" function in Equation 2.32 finds the integer number of the argument in parentheses by rounding downward. The "Mod(Y,4)" function finds the modulus (remainder) of the quotient Y/4.

For monthly periods, the day of the year at the middle of the month (J_{month}) is approximately:

$$J_{month} = Int(30.4\ M - 15) \qquad 2.38b$$

The sunset hour angle, ω_s, is given by:

$$\omega_s = \arccos\left[-\tan(\varphi)\tan(\delta)\right] \qquad 2.39$$

The "arccos" function is the arc-cosine function and represents the inverse of the cosine.

For hourly time periods, the solar time angle at the beginning and end of the period serve as integration endpoints for calculating R_a:

$$R_a = \frac{12}{\pi} G_{sc} d_r [(\omega_2 - \omega_1)\sin(\varphi)\sin(\delta) + \cos(\varphi)\cos(\delta)(\sin(\omega_2) - \sin(\omega_1))] \quad 2.40$$

where R_a has units of [MJ m^{-2} h^{-1}], G_{sc} is solar constant [4.92 MJ m^{-2} h^{-1}], ω_1 is the solar time angle at beginning of period [radians], and ω_2 is the solar time angle at end of period [radians].

ω_1 and ω_2 are given by:

$$\omega_1 = \omega - \frac{\pi t_1}{24} \quad 2.41$$

$$\omega_2 = \omega + \frac{\pi t_1}{24} \quad 2.42$$

where ω is solar time angle at the midpoint of the period in radians, and t_1 is the length of the calculation period in h: i.e., 1 for hourly periods or 0.5 for 30-min periods. The solar time angle at the midpoint of the hourly or shorter period is:

$$\omega = \frac{\pi}{12}\left[(t + 0.06667\,(L_z - L_m) + S_c) - 12\right] \quad 2.43$$

where t is standard clock time at the midpoint of the period in h (after correcting time for any daylight savings shift). For example, for a period between 1400 and 1500 h, t = 14.5 h, L_z is longitude of the center of the local time zone [expressed as positive degrees west of Greenwich, England (note that this sign convention is not congruent with common European usage). In the United States, L_z = 75°, 90°, 105° and 120° (west) for the Eastern, Central, Rocky Mountain and Pacific time zones, respectively, and L_z = 0° for Greenwich, 345° for Paris (France), and 255° for Bangkok (Thailand), L_m is the longitude of the solar radiation measurement site [expressed as positive degrees west of Greenwich, England], and S_c is a seasonal correction for solar time [hour].

Because ω_s is the sunset hour angle and $-\omega_s$ is the sunrise hour angle (noon has $\omega = 0$), values of $\omega < -\omega_s$ or $\omega > \omega_s$ from Equation 2.43 indicate that the sun is below the horizon, so that, by definition, R_a and R_{so} are zero and their calculation has no meaning. When the values for ω_1 and ω_2 span the value for $-\omega_s$ or for ω_s, this indicates that sunrise or sunset occurs within the hourly (or shorter) period. In this case, the integration limits for applying Equation 2.40 should be correctly set using the following conditionals:

If $\omega_1 < -\omega_s$ then $\omega_1 = -\omega_s$

If $\omega_2 < -\omega_s$ then $\omega_2 = -\omega_s$ 2.44

If $\omega_1 > \omega_s$ then $\omega_1 = \omega_s$

If $\omega_2 > \omega_s$ then $\omega_2 = \omega_s$

If $\omega_1 > \omega_2$ then $\omega_1 = \omega_2$

The above conditions can be applied for all time steps to insure numerical stability of the application of Equation 2.40 as well as correctly computing the theoretical quantity of solar radiation early and late in the day. Where there are hills or mountains, the hour angle when the sun first appears or disappears may increase for sunrise or decrease for sunset. The seasonal correction for solar time is:

$$S_c = 0.1645 \sin(2\,b) - 0.1255 \cos(b) - 0.025 \sin(b) \qquad 2.45$$

$$b = \frac{2\,\pi\,(J-81)}{364} \qquad 2.46$$

where J is the number of the day in the year and b has units of radians.

The user should confirm accurate setting of the data logger clock. If clock times are in error by more than 5–10 min, estimates of exoatmospheric and clear sky radiation may be significantly impacted. This can lead to errors in estimating R_n on an hourly or shorter basis, especially early and late in the day. A shift in "phase" between measured R_s and R_{so} estimated from R_a according to the data logger clock can indicate error in the reported time. More discussion is given in Appendix D of ASCE-EWRI (2005).

The angle of the sun above the horizon, β, at the midpoint of the hourly or shorter time period is computed as:

$$\beta = \arcsin\left[\sin(\phi)\sin(\delta) + \cos(\phi)\cos(\delta)\cos(\omega)\right] \qquad 2.47$$

where β has units of radians, φ is station latitude in radians, δ is solar declination in radians, and φ is solar time angle at the midpoint of the period in radians. The "arcsin" is the arc-sine function and represents the inverse of the sine.

2.3.3.10 Soil Heat Flux

According to the FAO-56 and ASCE-EWRI (2005), G is positive when the soil is warming and negative when the soil is cooling. For daily periods, the magnitude of G averaged over 24 hours beneath a fully vegetated grass or alfalfa reference surface is relatively small in comparison with R_n. Therefore, it is ignored in the standardized ET calculations so that:

$$G_{day} = 0 \qquad 2.48$$

where G_{day} is the daily (24-h) soil heat flux density [MJ m^{-2} day^{-1}].

Over a monthly period, G for the soil profile can be significant, especially during spring and fall. Assuming a constant soil heat capacity of 2.0 MJ m^{-3} °C^{-1}, and an effectively warmed soil depth of 2 m, G for monthly periods in MJ m^{-2} day^{-1} is estimated from the change in mean monthly air temperature as:

$$G_{month,i} = 0.07(T_{month,i+1} - T_{month,i-1}) \tag{2.49}$$

or, if $T_{month,i+1}$ is unknown:

$$G_{month,i} = 0.14(T_{month,i} - T_{month,i-1}) \tag{2.50}$$

where $T_{month,i}$ is mean air temperature of month i in °C, $T_{month,i-1}$ is mean air temperature of the previous month in °C, and $T_{month,i+1}$ is the mean air temperature of the next month in °C.

For hourly or shorter time periods, G, in the FAO-56 and ASCE-EWRI standardizations, is expressed as a function of net radiation for the two reference types. For the standardized short reference ET_0:

$$G_{hr\,daytime} = 0.1 R_n \tag{2.51a}$$

$$G_{hr\,nighttime} = 0.5 R_n \tag{2.51b}$$

where G and R_n have the same measurement units (MJ m^{-2} h^{-1} for hourly or shorter time periods). For standardization, nighttime is defined as when measured or calculated hourly net radiation R_n is < 0 (i.e., negative). The amount of energy consumed by G is subtracted from R_n when estimating ET_0. The coefficient 0.1 in Equation 2.51a represents the condition of only a small amount of dead thatch underneath the leaf canopy of the short (clipped grass) reference. Large amounts of thatch insulate the soil surface, reducing the daytime coefficient for grass to about 0.02. However, the 0.1 coefficient is part of the ASCE-EWRI (2005) and FAO-56 (Allen et al. 1998) standardizations.

2.3.4 *Limited Data Availability and Weather Data Integrity*

When calculating the Penman-Monteith ET_0, missing or poor quality data can be estimated using procedures described in FAO-56 (Allen et al. 1998) and in ASCE-EWRI (2005), or alternatively, the Hargreaves-Samani ET_0 equation (Hargreaves and Samani 1982; Hargreaves et al. 1985) can be applied. Hargreaves and Allen (2003) found similar accuracy between using the Hargreaves-Samani equation and the standardized Penman-Monteith method when only daily maximum and minimum air temperature were available. Weather data should be quality checked to insure integrity and representativeness. This is especially important with electronically collected data, since human oversight and maintenance may be limited. Solar radiation can be checked by plotting the measurements against a clear sky R_{so} envelope provided by Equation 2.34 or a more accurate and complicated method described in Appendix D of ASCE-EWRI (2005). Humidity data (T_d, RH, e_a) can be evaluated by examining daily maximum RH (RH_{max}) or by comparing T_d with T_{min}. Under reference conditions, RH_{max} generally approaches 100% during early morning and T_d approaches T_{min} (Allen 1996a; ASCE-EWRI 2005).

Weather data should be representative of the reference condition. Data collected at or near airports can be negatively influenced by the local aridity, especially in arid and semiarid climates. Data from dry or urban settings may cause overestimation of ET_0 due to air temperature measurements that are too high and humidity measurements that are too low, relative to the reference condition. Allen et al. (1998) and ASCE-EWRI (2005) suggest simple adjustments for "non-reference" weather data to provide data more reflective of well-watered settings. Allen and Gichuki (1989) and Ley et al. (1996) suggested more complicated approaches.

Often, substituting $T_d = T_{min} - K_o$ for measured T_d, as suggested in Equation 2.22, can improve ET_0 estimates made with the combination equation when data are from a non-reference setting. Using non-reference (i.e., arid) data in Equation 2.22 will tend to overestimate the true T_d and e_a that would occur under reference conditions, since T_{min} will be higher in the dry setting and consequently, so will estimated T_d. However, because e_s and e_a in the non-reference setting are both inflated when calculated using T_{max}, T_{min} and T_d estimated from Equation 2.27, the e_s - e_a difference in the combination equation is often brought more in line with that expected for the reference condition and a more accurate estimate for ET_0 results (ASCE-EWRI 2005).

2.4 The Hargreaves-Samani Method

As mentioned in Section 2.1, the most important reason to include a simpler method than the (default recommended) Penman-Monteith method is the likelihood for the lack of reliable meteorological data. The propagation of errors in the more difficult to measure parameters, such as radiation, humidity and wind speed, into the calculated $ET_{0,PM}$ value brought Hargreaves and Samani (1985), Hargreaves et al. (1985) and Hargreaves (1994) to derive an equation that needs mean daily maximum and mean daily minimum temperature as only input data:

$$ET_{0,har} = 0.0023 \times 0.408 R_A \times (T_{av} + 17.8) \times TD^{0.5} \qquad 2.52$$

where

R_A = extraterrestrial radiation (MJ/m²/day)

T_{av} = average daily air temperature (°C); $T_{av} = (T_{max} + T_{min})/2$
TD = difference between mean daily maximum and minimum temperature (°C); $TD = (T_{max} - T_{min})$

$ET_{0,har}$ has units of mm/day. The constant 0.408 converts radiation to evaporation equivalents in mm. The two parameters 0.0023 and 17.8 were obtained by Hargreaves and Samani (1985) and Hargreaves et al. (1985) by fitting measured ET_0 values to Equation 2.52. Droogers and Allen (2002) compared monthly values of $ET_{0,pm}$ with $ET_{0,har}$ using the global climate data set of the World Water and Climate Atlas (IWMI 2000). They developed a data set for 56,000 stations around the world and showed that Equation 2.52 tends to underestimate in the very dry regions and to overestimate in the very wet regions, but otherwise produces relatively reliable estimates relative to the standardized PM.

To reduce this difference, Droogers and Allen (2002) modified Equation 2.52 by adding precipitation. This parameter was selected because observations of precipitation are collected at a reasonable accuracy for a majority of the meteorological stations around the world, and with the assumption that monthly precipitation can in some regards represent relative levels of humidity. Also, precipitation data are needed to estimate the irrigation water requirements (Chapter 5). After testing various combinations based on Equation 2.52, the following equation was derived for application on monthly timesteps:

$$ET_{0,mh} = 0.0013 \times 0.408 R_A \times (T_{av} + 17.0) \times (TD - 0.0123P)^{0.76} \qquad 2.53$$

where, in addition to the above parameters, P is the precipitation (mm/month) and $ET_{0,mh}$ is monthly ET_0 in mm/day. This modified Hargreaves-Samani Method was better able to estimate ET_0 within the global data set. Figure 2.6 shows the related scatter plot of the difference between monthly ET_0 estimates using Penman-Monteith (PM) and modified Hargreaves-Samani (MH). A random 0.1% of the above 56,000 points for each month are plotted.

If the difference between mean daily maximum and minimum temperature (°C); $TD = (T_{max} - T_{min})$ is small while monthly precipitation (P) is high, the term ($TD - 0.0123P$) in Equation 2.53 could become negative (humid tropics). In such a case, the term cannot be raised to the power 0.76. To avoid this problem CRIWAR assumes $(TD - 0.0123P) = 0.1$ for all values of $(TD - 0.0123P) \leq 0.1$.

2.5 Discussion

As mentioned in Section 2.2, there is evidence that the FAO Modified Penman Method predicts a higher reference ET than the Penman-Monteith approach. With monthly average meteorological data from 20 stations, ET_{fao} and ET_{pm} have been calculated in Fig. 2.7 with Equations 2.13 and 2.14, respectively. A plot of the results in Fig. 2.7 shows $ET_{pm} = 0.85ET_{fao}$.

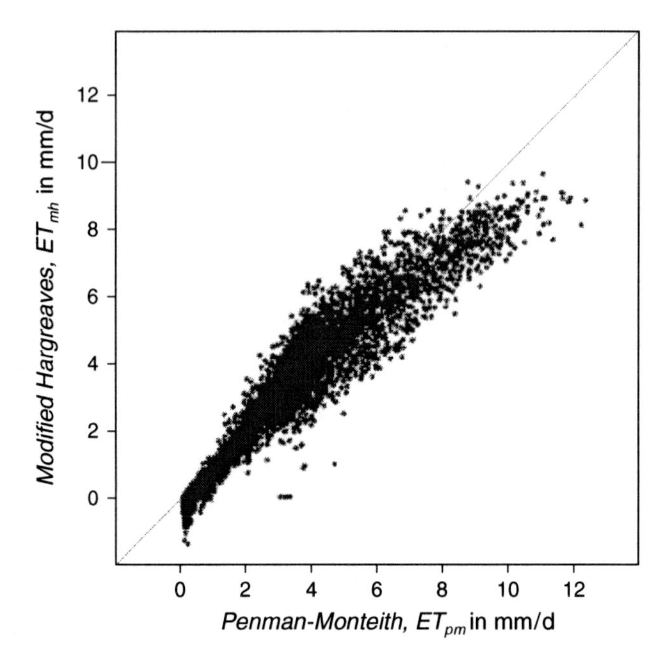

Fig. 2.6 Scatter plot of the difference between monthly ET_0 estimates using Penman-Monteith (PM) and modified Hargreaves-Samani (MH). A random 0.1% of the total points for each month are plotted (Droogers and Allen 2002)

As mentioned before, the modified Hargreaves-Samani Method estimates the reference *ET* as calculated with the Penman-Monteith equation while using less meteorological data. Thus, if the collected meteorological data contain significant error in solar radiation, humidity or wind speed, less of these errors will propagate into the estimate of ET_0 using the modified Hargreaves-Samani method. Obviously, no information is available on the error in collected data under such data-scarce conditions. Droogers and Allen (2002), however, discussed errors in data with experts having an extensive experience in observing meteorological data, especially in developing countries. Their estimate of the error with a 95% confidence interval (two times the standard deviation, assuming a normal distribution) are shown in Table 2.4.

The results in Table 2.4 show that for situations where the error in measured meteorological data is high, it may be better to use the Hargreaves-Samani method than to repair the inaccurate data set.

It should be noted, however, the Penman-Monteith method is the recommended (default) method for computing ET_0, if the accuracy of collected meteorological data is good. This especially because the Hargreaves-Samani method is a regression function derived from the Penman-Monteith method.

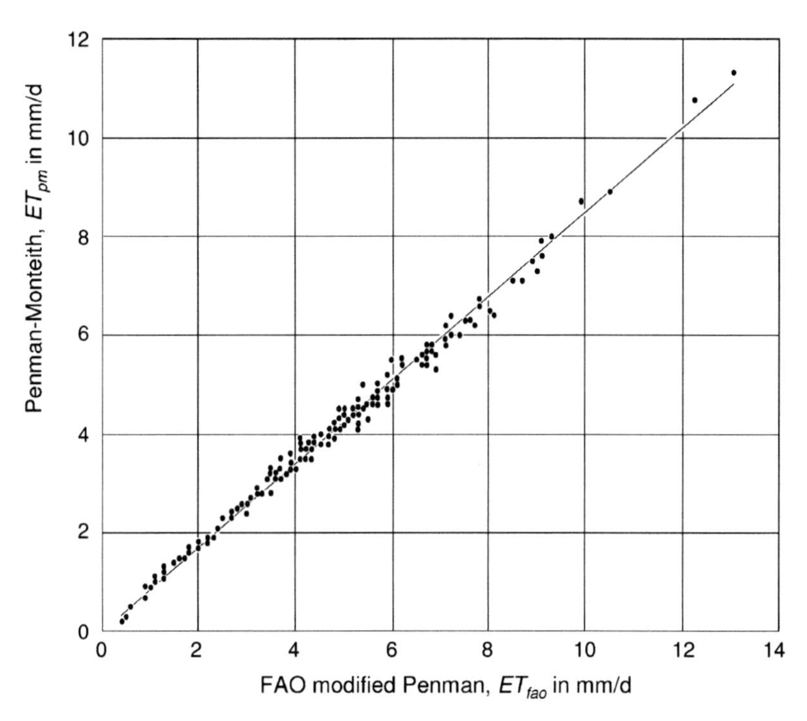

Fig. 2.7 Comparison of the reference ET for 20 locations, computed with the FAO modified Penman method (Equation 2.13) and the Penman-Monteith approach (Equation 2.14)

2.6 The Potential ET

The crop coefficient, K_c, has been developed over the past half-century to simplify and standardize the calculation and estimation of crop water use. The K_c is defined as the potential ratio of *ET* from a specific surface to ET_0. For crops, the potential ET (or ET_p) is assumed to be the evapotranspiration of a crop that has no reduction in transpiration due to soil water deficits. The specific surface can be comprised of bare soil, of soil with partial vegetation cover, or of full vegetation cover. The K_c represents an integration of effects of crop height, crop-soil resistance and surface albedo that distinguish the surface from the ET_0 definition. The value for K_c changes during the growing season as plants grow and develop, as the fraction of ground covered by vegetation changes, as the wetness of the underlying soil surface changes, and as plants age and mature. The potential *ET* is calculated by multiplying ET_0 by the crop coefficient:

$$ET_p = K_c \, ET_0 \qquad 2.54$$

The crop, and reference crop, are both living vegetative surfaces that are affected by variable weather. Differences between ET_0 and ET_p result mainly from differences in

Table 2.4 Effect of data measurement errors on Penman-Monteith and Hargreaves-Samani as compared with the standard global average ET_0 as calculated with Penman-Monteith without the shown measurement errors (Droogers and Allen 2002)

Data needed	Penman-Monteith	Hargreaves-Samani	Error with 95% confidence interval
Minimum temperature	✓	✓	5% (≈ 1°C)
Maximum temperature	✓	✓	5% (≈ 1°C)
Humidity	✓		25%
Wind speed	✓		25%
Radiation	✓		25%
Precipitation		✓	10%
R^2	0.871	0.915	
Root mean square difference	0.93 mm/day	0.72 mm/day	
Global daily average ET_0	3.00 mm/day	2.90 mm/day	

net radiation, stomatal effects on canopy resistance, and canopy height and roughness effects on aerodynamic resistance. The differences in net radiation are relatively fixed for any given growth stage unless the crop is on sloping topography, so most changes in the $K_c = ET_p/ET_0$ result from variations in stomatal control by the plants and the aerodynamic resistance, which depends on the canopy height and roughness and the wind speed. Therefore, we expect that K_c values are rather robust from one location to another unless the location has undulating terrain, a different climate has an effect on stomatal functions, or the wind speed is substantially different. Thus, one can generally transfer K_c values between locations with different climates. The ability to transfer has led to the widespread acceptance and usefulness of the $K_c \times ET_0$ approach.

As shown later in Section 2.6.1, estimating K_c is not a trouble-free task because it requires information on the vegetation status that is not always available. If optical satellite remote sensing is available, this can give accurate data on the vegetation status. Usually information is obtained through the vegetation index (VI). Reflectance values for the red (ρ_r) and the near infrared (ρ_{nir}) bands (0.6–0.7 µm and 0.7–1.3 µm respectively) are used to calculate the Normalized Difference Vegetation Index (NDVI):

$$NDVI = \frac{\rho_{nir} - \rho_r}{\rho_{nir} + \rho_r} \qquad 2.55$$

A linear relationship between the NDVI and the K_c was introduced by Heilman et al. (1982) and theoretically established by Choudhury (1994). The resulting equation is:

$$K_c = 1.25 \times NDVI + 0.2 \qquad 2.56$$

Tasumi et al. (2005) found a similar relationship. NDVI values can be taken from all high-resolution satellite images for all crops (land uses) over the gross command area. NDVI values from different satellites show close correlation (Belmonte et al. 2005; Fig. 2.8).

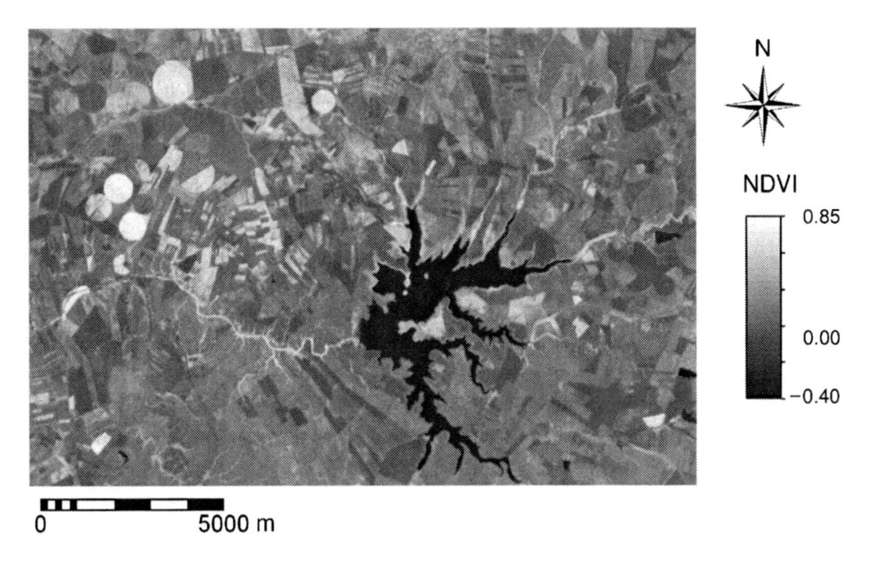

Fig. 2.8 NDVI for the Roxo reservoir and the downstream irrigation scheme. Date: 25 May 2003 Landsat 7 bands 3 and 4. Place: Beja, Portugal 10h 57 min UTC, 38° 01′ 7° 52′ W (Courtesy Ambro Gieske, ITC)

2.6.1 The Crop Coefficient Curve

The crop coefficient K_c is defined as the ratio of ET_p from the specific crop to ET_0 from the reference crop. The crop coefficient curve represents the changes in K_c over the course of the growing season, depending on changes in vegetation cover and physiology. During the initial period, shortly after planting of annuals or prior to the initiation of new leaves for perennials, the value of K_c is small, often less than 0.4.

The simple, linear K_c approximation between critical growth points was proposed by FAO in Doorenbos and Pruitt (1977) and Allen et al. (1998). The method is still widely used and generally provides sufficiently accurate descriptions of the annual K_c curve for most applications. Definitions for three benchmark K_c values required to construct the curve and associated definitions for growth stage periods and relative ground cover are illustrated in Fig. 2.9.

The linear K_c approximation curve is constructed by the following steps:

1. Divide the growing period into four general growth stages describing crop canopy development and phenology for a regional specific developmental crop calendar (Fig. 2.9). The four stages are (1) *Initial period* (planting or green-up until about 10% ground cover); (2) *Crop Development period*; from 10% ground cover until about 70% ground cover and higher; (3) *Mid Season period*; from 70% ground cover to the beginning of the late season period (the onset of senescence); and (4) *Late Season period* (beginning of senescence or mid grain or fruit fill until harvest, crop death, frost-kill, or full senescence).

2. Specify the three K_c values corresponding to $K_{c\ ini}$, $K_{c\ mid}$ and $K_{c\ end}$, where $K_{c\ ini}$ represents the average K_c during the initial period, $K_{c\ mid}$ represents the average K_c during the midseason period and $K_{c\ end}$ represents the K_c at the end of the late season period.
3. Connect straight line segments through each of the four growth stage periods, with horizontal lines drawn through $K_{c\ ini}$ during the initial period and through $K_{c\ mid}$ during the midseason period. Diagonal lines are drawn from $K_{c\ ini}$ to $K_{c\ mid}$ within the domain of the development period and from $K_{c\ mid}$ to $K_{c\ end}$ within the domain of the late season period.

Table 2.5 lists $K_{c\ ini}$, $K_{c\ mid}$ and $K_{c\ end}$ for a large number of crops. The three K_c columns represent typical irrigation management and precipitation frequencies. The majority of K_c values are taken from FAO-56 (Allen et al. 1998) which were largely based on Doorenbos and Pruitt (1977) and Doorenbos and Kassam (1979). K_c values in Table 2.5 for tree crops have been expanded from those in FAO-56 to show entries for various fractions of surface covered by vegetation ($f_{c\ eff}$) using Equations 2.69–2.71 that are introduced later. The K_c values in Table 2.5 are applicable with grass reference ET_0 as defined and represented by the standardized FAO/ASCE-PM Equation 2.14, and are generally valid for use with ET_{ref} by other grass reference equations, provided these agree with the standardized PM definition (Allen et al. 2005a).

The K_c values in Table 2.5 are organized by crop group type because there is similarity in K_c among the members of the same crop group due to similarity in

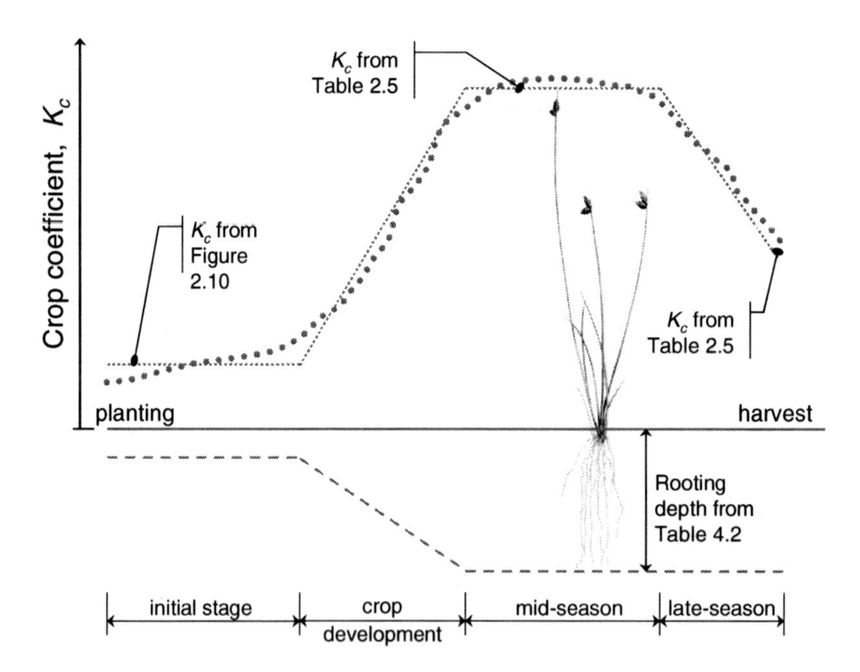

Fig. 2.9 FAO-style, linearized K_c curve with four crop stages and three K_c values relative to typical ground cover

plant height, leaf area, ground coverage, stomatal behavior, and water management. For several group types, only one value for $K_{c\,ini}$ is listed for the whole group, since tabularized $K_{c\,ini}$ are only approximate. Figure 2.10a–c from FAO-56 (Allen et al. 1998) discussed later improve estimates for $K_{c\,ini}$ by accounting for frequency of soil wetting and soil type (see Section 2.6.4).

2.6.2 The Climatic Basis of Table 2.5

The $K_{c\,mid}$ values in Table 2.5 are typical values expected under a standard climatic condition defined in FAO-56 (Allen et al. 1998) as a sub-humid climate having average daytime minimum relative humidity (RH_{min} = 45%) and having calm to moderate wind speeds averaging 2 m s^{-1}. More arid climates and conditions of greater wind speed have higher values for K_c, especially for tall crops, and more humid climates with lower wind speed reduce the K_c values according to the relationship:

$$K_c = K_{ctable} + \left[0.04\left(u_2 - 2\right) - 0.004\left(RH_{min} - 45\right)\right]\left(\frac{h}{3}\right)^{0.3} \qquad 2.57$$

where $K_{c\,table}$ is the K_c value (or K_{cb} value) from Table 2.5 for $K_{c\,mid}$ (and for $K_{c\,end}$ when $K_{c\,end} > 0.45$) and h is mean crop height in medium The K_c for crops between 2 and 3 m in height can increase by as much as 40% when going from a calm, humid climate (for example, u_2 = 1 m s^{-1} and RH_{min} = 70%) to an extremely windy, arid climate (for example, u_2 = 5 m s^{-1} and RH_{min} = 15%). The increase in K_c is due to the influence of the larger aerodynamic roughness of tall crops relative to grass on the transport of water vapor from the surface. The adjustments to K_c for climate are generally made using mean values for u_2 and RH_{min} during the midseason period. Typical values for h are included in Tables 2.7 and 4.2.

2.6.3 Lengths of Crop Growth Stages

The four crop growth stages for the FAO-56 (Allen et al. 1998) K_c curves are characterized in terms of crop growth benchmarks as illustrated in Fig. 2.8. The crop development period is presumed to begin when approximately 10% of the ground is covered by vegetation and ends at attainment of effective full cover, which typically occurs at 70% or more ground covered (shaded) by canopy. Effective cover can be defined for row crops such as beans, sugar beets, potatoes and corn, as the time when some leaves of plants in adjacent rows begin to intermingle so that soil shading becomes nearly complete, or when plants reach nearly full size, if no intermingling occurs. For crops taller than 0.5 m, the average fraction of the ground

Table 2.5 Mean crop coefficients, K_c, and basal crop coefficients, K_{cb}, for well-managed crops in a sub-humid climate for use with ET_0 (After FAO-56*; Allen et al. 1998)

Crop	$K_{c\ ini}$ [1]	$K_{c\ mid}$	$K_{c\ end}$	$K_{cb\ ini}$	$K_{cb\ mid}$	$K_{cb\ end}$
a. Small vegetables	0.7	1.05	0.95	0.15	0.95	0.85
Brócoli		1.05	0.95		0.95	0.85
Brussels Sprouts		1.05	0.95		0.95	0.85
Cabbage		1.05	0.95		0.95	0.85
Carrots		1.05	0.95		0.95	0.85
Cauliflower		1.05	0.95		0.95	0.85
Celery		1.05	1.00		0.95	0.90
Garlic		1.00	0.70		0.90	0.60
Lettuce		1.00	0.95		0.90	0.90
Onions – dry		1.05	0.75		0.95	0.65
– green		1.00	1.00		0.90	0.90
– seed		1.05	0.80		1.05	0.70
Spinach		1.00	0.95		0.90	0.85
Radish		0.90	0.85		0.85	0.75
b. Vegetables – Solanum Family (*Solanaceae*)	0.6	1.15	0.80	0.15	1.10	0.70
Egg Plant		1.05	0.90		1.00	0.80
Sweet Peppers (bell)		1.05[2]	0.90		1.00[2]	0.80
Tomato		1.15	0.70–0.90		1.10[2]	0.60–0.80
c. Vegetables – Cucumber Family (*Cucurbitaceae*)	0.5	1.00	0.80	0.15	0.95	0.70
Cantaloupe	0.5	0.85	0.60		0.75	0.50
Cucumber – fresh market	0.6	1.00[2]	0.75		0.95[2]	0.70
– machine harvest	0.5	1.00	0.90		0.95	0.80
Pumpkin, Winter Squash		1.00	0.80		0.95	0.70
Squash, Zucchini		0.95	0.75		0.90	0.70
Sweet Melons		1.05	0.75		1.00	0.70
Watermelon	0.4	1.00	0.75		0.95	0.70
d. Roots and Tubers	0.5	1.10	0.95	0.15	1.00	0.85
Beets, table		1.05	0.95		0.95	0.85
Cassava – year 1	0.3	0.80[3]	0.30		0.70[3]	0.20
– year 2	0.3	1.10	0.50		1.00	0.45
Parsnip	0.5	1.05	0.95		0.95	0.85
Potato		1.15	0.75[4]		1.10	0.65[4]
Sweet Potato		1.15	0.65		1.10	0.55
Turnip (and Rutabaga)		1.10	0.95		1.00	0.85
Sugar Beet	0.35	1.20	0.70[5]		1.15	0.50[5]
e. Legumes (*Leguminosae*)	0.4	1.15	0.55	0.15	1.10	0.50
Beans, green	0.5	1.05[2]	0.90		1.00[2]	0.80
Beans, dry and Pulses	0.4	1.15[2]	0.35		1.10[2]	0.25
Chick pea		1.00	0.35		0.95	0.25
Fababean (broad bean) – fresh	0.5	1.15[2]	1.10		1.10[2]	1.05
– dry/seed	0.5	1.15[2]	0.30		1.10[2]	0.20
Garbanzo	0.4	1.15	0.35		1.05	0.25
Green Gram and Cowpeas		1.05	0.60–0.35[6]		1.00	0.55–0.25[6]
Groundnut (Peanut)		1.15	0.60		1.10	0.50

(continued)

Table 2.5 (continued)

Crop	$K_{c\ ini}$ [1]	$K_{c\ mid}$	$K_{c\ end}$	$K_{cb\ ini}$	$K_{cb\ mid}$	$K_{cb\ end}$
Lentil		1.10	0.30		1.05	0.20
Peas – fresh	0.5	1.15[2]	1.10		1.10[2]	1.05
– dry/seed		1.15	0.30		1.10	0.20
Soybeans		1.15	0.50		1.10	0.30
f. Perennial Vegetables (with winter dormancy and initially bare or mulched)	0.5	1.00	0.80			
Artichokes	0.5	1.00	0.95	0.15	0.95	0.90
Asparagus	0.5	0.95[7]	0.30	0.15	0.90[7]	0.20
Mint	0.60	1.15	1.10	0.40	1.10	1.05
Strawberries	0.40	0.85	0.75	0.30	0.80	0.70
g. Fiber Crops	0.35			0.15		
Cotton		1.15–1.20	0.70–0.50		1.10–1.15	0.50–0.40
Flax		1.10	0.25		1.05	0.20
Sisal[8]		0.4–0.7	0.4–0.7		0.4–0.7	0.4–0.7
h. Oil Crops	0.35	1.15	0.35	0.15	1.10	0.25
Castorbean (*Ricinus*)		1.15	0.55		1.10	0.45
Rapeseed, Canola		1.0–1.15[9]	0.35		0.95–1.10[9]	0.25
Safflower		1.0–1.15[9]	0.25		0.95–1.10[9]	0.20
Sesame		1.10	0.25		1.05	0.20
Sunflower		1.0–1.15[9]	0.35		0.95–1.10[9]	0.25
i. Cereals	0.3	1.15	0.4	0.15	1.10	0.25
Barley		1.15	0.25		1.10	0.15
Oats		1.15	0.25		1.10	0.15
Spring Wheat		1.15	0.25–0.4[10]		1.10	0.15–0.3[10]
Winter Wheat – with frozen soils	0.4	1.15	0.25–0.4[10]	0.15–0.5[11]	1.10	0.15–0.310
– with non-frozen soils	0.7	1.15	0.25–0.4[10]			
Maize, Field (grain) (*field corn*)		1.25[12]	0.60,0.35[13]	0.15	1.20[12]	0.50, 0.15[13]
Maize, Sweet (*sweet corn*)		1.15[12]	1.05[14]		1.10[12]	1.00[14]
Millet		1.00	0.30		0.95	0.20
Sorghum – grain		1.00–1.10	0.55		0.95–1.05	0.35
– sweet		1.20	1.05		1.15	1.00
Rice	1.05	1.05–1.20[15]	0.90–0.60	1.00	1.00–1.15[15]	0.70–0.45
j. Forages						
Alfalfa Hay – averaged cutting effects	0.40	0.95[16]	0.90			
– individual cutting periods	0.40[17]	1.20[17]	1.15[17]	0.30[17]	1.15[17]	1.10[17]
– for seed	0.40	0.50	0.50	0.30	0.45	0.45
Bermuda Hay – averaged cutting effects	0.55	1.00[16]	0.85	0.50	0.95[18]	0.80
– spring crop for seed	0.35	0.90	0.65	0.15	0.85	0.60
Clover Hay, Berseem						
– averaged cutting effects	0.40	0.90[16]	0.85			
– individual cutting periods	0.40[17]	1.15[17]	1.10[17]	0.30[17]	1.10[17]	1.05[17]

(continued)

Table 2.5 (continued)

Crop	$K_{c\ ini}$[1]	$K_{c\ mid}$	$K_{c\ end}$	$K_{cb\ ini}$	$K_{cb\ mid}$	$K_{cb\ end}$
Rye Grass Hay – averaged cutting effects	0.95	1.05	1.00	0.85	1.00[18]	0.95
Sudan Grass Hay (annual)						
– average cutting effects	0.50	0.90[17]	0.85			
–individual cutting periods	0.50[17]	1.15[17]	1.10[17]	0.30[17]	1.10[17]	1.05[17]
Grazing Pasture – rotated grazing	0.40	0.85–1.05	0.85	0.30	0.80–1.00	0.80
– extensive grazing	0.30	0.75	0.75	0.30	0.70	0.70
Switchgrass[19]	0.20	1.05	0.20	0.15	1.00	0.10
Turf grass – cool season[20]	0.90	0.90	0.90	0.80	0.85	0.85
– warm season[20]	0.85	0.90	0.90	0.75	0.80	0.80
k. Sugar Cane	0.40	1.25	0.75	0.15	1.20	0.70
l. Tropical Fruits and Trees						
Banana – 1st year	0.50	1.10	1.00	0.15	1.05	0.90
– 2nd year	1.00	1.20	1.10	0.60	1.10	1.05
Cacao	1.00	1.05	1.05	0.90	1.00	1.00
Coffee – bare ground cover	0.90	0.95	0.95	0.80	0.90	0.90
– with weeds	1.05	1.10	1.10	1.00	1.05	1.05
Palms (including date palms)[21]						
– no ground cover	0.90	0.95	0.95	0.80	0.85	0.85
– high density (f_{ceff} = 0.7)[22]						
– no ground cover	0.80	0.80	0.80	0.70	0.70	0.70
– medium density (f_{ceff} = 0.5)						
– no ground cover	0.50	0.55	0.55	0.40	0.45	0.45
– low density/young (f_{ceff} = 0.25)						
– no ground cover	0.35	0.35	0.35	0.25	0.25	0.25
– very low density/young (f_{ceff} = 0.1)						
– active ground cover[31]	0.95	0.95	0.95	0.85	0.90	0.90
– high density (f_{ceff} = 0.7)						
– active ground cover	0.90	0.90	0.90	0.80	0.85	0.85
– medium density (f_{ceff} = 0.5)						
– active ground cover	0.85	0.85	0.85	0.75	0.80	0.80
– low density/young (f_{ceff} = 0.25)						
– active ground cover	0.80	0.80	0.80	0.70	0.75	0.75
– very low density/young (f_{ceff} = 0.1)						
Pineapple[23] – bare soil	0.50	0.30	0.30	0.15	0.25	0.25
– with grass cover	0.50	0.50	0.50	0.30	0.45	0.45
Rubber Trees	0.95	1.00	1.00	0.85	0.90	0.90
Tea – non-shaded	0.95	1.00	1.00	0.90	0.95	0.90
– shaded[24]	1.10	1.15	1.15	1.00	1.10	1.05
m. Grapes and Berries						
Berries (bushes)	0.30	1.05	0.50	0.20	1.00	0.40
Grapes – Table or Raisin[21]						

(continued)

Table 2.5 (continued)

Crop	$K_{c\ in}$	$K_{c\ mid}$	$K_{c\ end}$	$K_{cb\ ini}$	$K_{cb\ mid}$	$K_{cb\ en}$
– no ground cover	0.29	1.09	0.87[27]	0.19	1.04	0.82[27]
– high density (f_{ceff} = 0.7)[25]						
– no ground cover	0.29	0.95	0.76[27]	0.19	0.90	0.71[27]
– medium density (f_{ceff} = 0.5)[22]						
– no ground cover	0.27	0.58	0.48[27]	0.17	0.53	0.43[27]
– low/young (f_{ceff} = 0.25)						
Grapes – Wine						
– no ground cover	0.30	0.75[23]	0.60[26, 27]	0.20	0.70[26]	0.55[26, 27]
– high density (f_{ceff} = 0.7)						
– no ground cover	0.30	0.70[23]	0.55[26, 27]	0.20	0.65[26]	0.50[26, 27]
– medium density (f_{ceff} = 0.5)[22]						
– no ground cover	0.30	0.45[23]	0.40[26, 27]	0.25	0.40[26]	0.30[26, 27]
– low/young (f_{ceff} = 0.25)						
Hops	0.3	1.05	0.85	0.15	1.00	0.80
n. Fruit Trees						
Almonds[21]						
– no ground cover	0.40	1.00	0.70[27]	0.20	0.95	0.65[27]
– high density (f_{ceff} = 0.7)						
– no ground cover	0.40	0.85	0.60[27]	0.20	0.80	0.55[27]
– medium density (f_{ceff} = 0.5)[22]						
– no ground cover	0.35	0.50	0.40[27]	0.15	0.45	0.35[27]
– low density/young (f_{ceff} = 0.25)						
– active ground cover[33]	0.85	1.05	0.85[27]	0.75	1.00	0.80[27]
– high density (f_{ceff} = 0.7)						
– active ground cover	0.85	1.00	0.85[27]	0.75	0.95	0.80[27]
– low/young (f_{ceff} = 0.25)						
– medium density (f_{ceff} = 0.5)						
– active ground cover	0.85	0.95	0.85[27]	0.75	0.90	0.80[27]
– low density/young (f_{ceff} = 0.25)						
Apples, Cherries, Pears[21]						
– no ground cover	0.50	1.15	0.80[27]	0.30	1.10	0.75[27]
– high density (f_{ceff} = 0.7)						
– no ground cover	0.45	1.05	0.75[27]	0.30	1.00[28]	0.70[27]
– medium density (f_{ceff} = 0.5)[22]						
– no ground cover	0.40	0.70	0.55[27]	0.25	0.65	0.50[27]
– low density/young (f_{ceff} = 0.25)						
– active ground cover,[33] killing frost	0.50	1.20	0.85[27]	0.40	1.15	0.80[27]
– high density (f_{ceff} = 0.7)						
– active ground cover, killing frost	0.50	1.15	0.85[27]	0.40	1.10	0.80[27]
– medium density (f_{ceff} = 0.5)[22]						

(continued)

Table 2.5 (continued)

Crop	$K_{c\ ini}$ [1]	$K_{c\ mid}$	$K_{c\ end}$	$K_{cb\ ini}$	$K_{cb\ mid}$	$K_{cb\ end}$
– active ground cover, killing frost	0.50	1.05	0.85[27]	0.40	1.00	0.80[27]
– low density (f_{ceff} = 0.25)						
– active ground cover, no frosts	0.85	1.20	0.85[27]	0.75	1.15	0.80[27]
– high density (f_{ceff} = 0.7)						
– active ground cover, no frosts	0.85	1.15	0.85[27]	0.75	1.10	0.80[27]
– medium density (f_{ceff} = 0.5)[22]						
– active ground cover, no frosts	0.85	1.05	0.85[27]	0.75	1.00	0.80[27]
– low density (f_{ceff} = 0.25)						
Apricots, Peaches, Stone Fruit[21, 29]						
– no ground cover	0.50	1.20	0.85[27]	0.30	1.15	0.80[27]
– super density (f_{ceff} = 0.9)[25]						
– no ground cover	0.50	1.15	0.80[27]	0.30	1.10	0.75[27]
– high density (f_{ceff} = 0.7)[30]						
– no ground cover	0.45	1.0	0.70[27]	0.25	0.95	0.65[27]
– medium density (f_{ceff} = 0.5)[22]						
– no ground cover	0.40	0.60	0.45[27]	0.20	0.55	0.40[27]
– low density/young (f_{ceff} = 0.25)[31]						
– active ground cover,[33] killing frost	0.50	1.25	0.85[27]	0.40	1.20	0.80[27]
– super density (f_{ceff} = 0.9)						
– active ground cover, killing frost	0.50	1.20	0.85[27]	0.40	1.15	0.80[27]
– high density (f_{ceff} = 0.7)[22]						
– active ground cover, killing frost	0.50	1.15	0.85[27]	0.40	1.10	0.80[27]
– medium density (f_{ceff} = 0.5)						
– active ground cover, killing frost	0.50	1.00	0.85[27]	0.40	0.95	0.80[27]
– low density (f_{ceff} = 0.25)						
– active ground cover, no frosts	0.80	1.25	0.85[27]	0.70	1.20	0.80[27]
– super density (f_{ceff} = 0.9)						
– active ground cover, no frosts	0.80	1.20	0.85[27]	0.70	1.15	0.80[27]
– high density (f_{ceff} = 0.7)[22]						
– low/young (f_{ceff} = 0.25)	0.80	1.15	0.85[27]	0.70	1.10	0.80[27]
– active ground cover, no frosts						
– medium density (f_{ceff} = 0.5)						
– active ground cover, no frosts	0.80	1.00	0.85[27]	0.70	0.95	0.80[27]
– low density (f_{ceff} = 0.25)						
Avocado[21]						
– no ground cover	0.50	1.00	0.90	0.30	0.95	0.85
– high density (f_{ceff} = 0.7)						
– no ground cover	0.50	0.90	0.80	0.30	0.85	0.80
– medium density (f_{ceff} = 0.5)[22]						

(continued)

Table 2.5 (continued)

Crop	$K_{c\ ini}$[1]	$K_{c\ mid}$	$K_{c\ end}$	$K_{cb\ ini}$	$K_{cb\ mid}$	$K_{cb\ end}$
– no ground cover – low density/young (f_{ceff} = 0.25)	0.40	0.65	0.60	0.25	0.60	0.50
– active ground cover[33] – high density (f_{ceff} = 0.7)	0.85	1.05	0.95	0.75	1.00	0.90
– active ground cover – medium density (f_{ceff} = 0.5)	0.85	1.00	0.95	0.75	0.95	0.90
– active ground cover – low density/young (f_{ceff} = 0.25)	0.85	0.95	0.90	0.75	0.90	0.85
Citrus[21]						
– no ground cover – high density (f_{ceff} = 0.7)[32]	0.95	0.90	0.90	0.85	0.85	0.85
– no ground cover – medium density (f_{ceff} = 0.5)	0.80	0.75	0.75	0.70	0.70	0.70
– no ground cover – low density/young (f_{ceff} = 0.25)	0.55	0.50	0.50	0.45	0.45	0.45
– active ground cover[33] – high density (f_{ceff} = 0.7)[34]	1.00	0.95	0.95	0.90	0.90	0.90
– active ground cover – medium density (f_{ceff} = 0.5)	0.95	0.95	0.95	0.85	0.90	0.90
– active ground cover – low density/young (f_{ceff} = 0.25)	0.90	0.90	0.90	0.80	0.85	0.85
Conifer Trees[35]	1.00	1.00	1.00	0.95	0.95	0.95
Kiwi	0.40	1.05	1.05	0.20	1.00	1.00
Mango[21]						
– no ground cover – high density (f_{ceff} = 0.7)[36]	0.35	0.90	0.75	0.25	0.85	0.70
– no ground cover – medium density (f_{ceff} = 0.5)	0.35	0.75	0.60	0.25	0.70	0.55
– no ground cover – low density/young (f_{ceff} = 0.25)	0.30	0.45	0.40	0.20	0.40	0.35
Olives[21]						
– no ground cover – high density (f_{ceff} = 0.7)[22, 37]	0.65	0.70	0.60	0.55	0.65	0.55
– no ground cover – medium density (f_{ceff} = 0.5)[38]	0.60	0.60	0.55	0.50	0.55	0.50
– no ground cover – low density/young (f_{ceff} = 0.25)[39]	0.40	0.40	0.35	0.30	0.35	0.30
– no ground cover – very low density/young (f_{ceff} = 0.05)[39]	0.30	0.25	0.25	0.20	0.20	0.20

(continued)

Table 2.5 (continued)

Crop	$K_{c\ ini}$ [1]	$K_{c\ mid}$	$K_{c\ end}$	$K_{cb\ ini}$	$K_{cb\ mid}$	$K_{cb\ end}$
– active ground cover[33] – high density (f_{ceff} = 0.7)	0.80	0.75	0.75	0.70	0.70	0.70
– active ground cover – medium density (f_{ceff} = 0.5)	0.80	0.75	0.75	0.70	0.70	0.70
– active ground cover – low density/young (f_{ceff} = 0.25)	0.80	0.75	0.75	0.70	0.70	0.70
– active ground cover – very low density/young (f_{ceff} = 0.05)	0.80	0.75	0.75	0.70	0.70	0.70
Pistachios[21]						
– no ground cover – high density (f_{ceff} = 0.7)	0.40	1.00	0.70	0.30	0.95	0.65
– no ground cover – medium density (f_{ceff} = 0.5)	0.35	0.85	0.60	0.25	0.80	0.55
– no ground cover – low density/young (f_{ceff} = 0.25)	0.30	0.50	0.40	0.20	0.45	0.35
– active ground cover[33] – high density (f_{ceff} = 0.7)	0.80	1.00	0.75	0.70	0.95	0.70
– active ground cover – medium density (f_{ceff} = 0.5)	0.80	1.00	0.75	0.70	0.95	0.70
– active ground cover – low density/young (f_{ceff} = 0.25)	0.80	0.85	0.75	0.70	0.80	0.70
Walnut Orchard[21]						
– no ground cover – high density (f_{ceff} = 0.7)[22]	0.50	1.10	0.65[27]	0.40	1.05	0.60[27]
– no ground cover – medium density (f_{ceff} = 0.5)	0.45	0.90	0.60[27]	0.35	0.85	0.55[27]
– no ground cover – low density/young (f_{ceff} = 0.25)	0.35	0.55	0.40[27]	0.25	0.50	0.35[27]
– active ground cover[33] – high density (f_{ceff} = 0.7)	0.85	1.15	0.85[27]	0.75	1.10	0.80[27]
– active ground cover – medium density (f_{ceff} = 0.5)	0.85	1.10	0.85[27]	0.75	1.05	0.80[27]
– active ground cover – low density/young (f_{ceff} = 0.25)	0.85	0.95	0.85[27]	0.75	0.90	0.80[27]
o. Wetlands – temperate climate						
Cattails, Bulrushes, killing frost	0.30	1.20	0.30			
Cattails, Bulrushes, no frost	0.60	1.20	0.60			
Short Veg., no frost	1.05	1.10	1.10			
Reed Swamp, standing water	1.00	1.20	1.00			
Reed Swamp, moist soil	0.90	1.20	0.70			

(continued)

Table 2.5 (continued)

Crop	$K_{c\ ini}$ [1]	$K_{c\ mid}$	$K_{c\ end}$	$K_{cb\ ini}$	$K_{cb\ mid}$	$K_{cb\ end}$
p. Special						
Open Water, <2 m depth or in sub-humid climates or tropics		1.05	1.05			
Open Water, >5 m depth, clear of turbidity, temperate climate		0.50–0.70[39]	0.80–1.30[39]			

[1] These are general values for $K_{c\ ini}$ under typical irrigation management and soil wetting. For frequent wettings such as with high frequency sprinkle irrigation or daily rainfall, these values may increase substantially and may approach 1.0 to 1.2. $K_{c\ ini}$ is a function of wetting interval and potential evaporation rate during the initial and development periods and is more accurately estimated using Fig. 2.3 or using the dual $K_{cb\ ini} + K_e$.

[2] Beans, Peas, Legumes, Tomatoes, Peppers and Cucumbers are sometimes grown on stalks reaching 1.5 –2 m in height. In such cases, increased K_c values need to be taken. For green beans, peppers and cucumbers, 1.15 can be taken, and for tomatoes, dry beans and peas, 1.20. Under these conditions h should be increased also.

[3] The midseason values for cassava assume non-stressed conditions during or following the rainy season. The $K_{c\ end}$ and $K_{cb\ end}$ values account for dormancy during the dry season.

[4] The $K_{c\ end}$ and $K_{cb\ end}$ values for potatoes are about 0.40 and 0.35 for long season potatoes with vine kill.

[5] These $K_{c\ end}$ and $K_{cb\ end}$ values are for no irrigation during the last month of the growing season. The $K_{c\ end}$ and $K_{cb\ end}$ values for sugar beets are higher, up to 1.0 and 0.9, when irrigation or significant rain occurs during the last month.

[*] Primary source: FAO-56 (Allen et al. 1998), with information from Doorenbos and Kassam (1979), Doorenbos and Pruitt (1977); Pruitt (1986); Wright (1981, 1982), Snyder et al. (1989a, b).

[6] The first $K_{c\ end}$ is for harvested fresh. The second value is for harvested dry.

[7] The K_c for asparagus usually remains at $K_{c\ ini}$ during harvest of the spears, due to sparse ground cover. The $K_{c\ mid}$ value is for following regrowth of plant vegetation following termination of harvest of spears.

[8] K_c for sisal depends on the planting density and water management (e.g., intentional moisture stress).

[9] The lower values are for rainfed crops having less dense plant populations.

[10] The higher value is for hand-harvested crops.

[11] The two $K_{cb\ ini}$ values for winter wheat are for less than 10% ground cover and for during the dormant, winter period, if the vegetation fully covers the ground, but conditions are non-frozen.

[12] These $K_{c\ mid}$ and $K_{cb\ mid}$ values for maize are for robust, pristine crops having plant populations of 50,000 plants per ha or higher. For less dense populations or uniform growth, $K_{c\ mid}$ and $K_{cb\ mid}$ can be reduced by 0.10 to 0.2.

[13] The first $K_{c\ end}$ value is for harvest at high grain moisture. The second $K_{c\ end}$ value is for harvest after complete field drying of the grain (to about 18% moisture, wet mass basis).

[14] If harvested fresh for human consumption. Use $K_{c\ end}$ for field maize if sweet maize is allowed to dry in the field.

[15] The low value for rice is for dense, uniform stands having low aerodynamic roughness (smooth canopy surface) and also low to moderate wind conditions (<2 m s^{-1}). The higher value is for somewhat more sparse, but inundated (flooded) conditions having greater roughness and lower albedo caused by shadowing, due to the sparseness.

[16] This $K_{c\ mid}$ value for hay crops is an overall average $K_{c\ mid}$ value that averages K_c before and following cuttings. It is applied to the period following the first development period until the beginning of the last late season period.

Table 2.5 (continued)

[17]These K_c values for hay crops represent immediately following cutting; at full cover; and immediately before cutting, respectively. The growing season is described as a series of individual cutting periods (Fig. 2.4).
[18]This $K_{cb\ mid}$ value for Bermuda and ryegrass hay is an overall average $K_{cb\ mid}$ value that averages K_{cb} before and following cuttings. It is applied to the period following the first development period until the beginning of the last late season period.
[19]Based on measurements of ET from prairie in Kansas by Verma et al. (1991) comprised of switchgrass (*Panicum virgatum*), big bluestem (*Andropogon gerardii*) and indiangrass (*Sorghastrum nutans*).
[20] Cool season grass varieties include dense stands of bluegrass, ryegrass, and fescue. Warm season varieties include bermuda grass and St. Augustine grass. The values given here are for potential conditions representing a 0.06–0.08 m mowing height. Turf, especially warm season varieties, can be stressed at moderate levels and still maintain appearance (see section "Evapotranspiration Coefficients for Landscapes" and Table 2.12). Generally a value for the stress coefficient K_s of 0.9 for cool season and 0.7 for warm season varieties can be employed where careful water management is practiced and rapid growth is not required (Table 2.12). Incorporation of these values for K_s into an 'actual K_c' using potential values in this table will yield $K_{c\ act}$ values of about 0.8 for cool season turf and 0.65 for warm season turf.
[21]These values for $K_{cb\ ini}$, $K_{cb\ mid}$ and $K_{cb\ end}$ were modeled using Equations 2.67, 2.72 and 2.73 and parameters lised in Table 2.3, along with $f_{c\ eff}$ and h from Table 2.4, where $f_{c\ eff}$ is the effective fraction of ground covered or shaded by vegetation (0 to 1.0) near solar noon and h is the mean height of the vegetation.
[22]The values in this row are similar to the entry in FAO-56 (Allen et al. 1998).
[23]The pineapple plant has very low transpiration because it closes its stomata during the day and opens them during the night. Therefore, the majority of ET_c from pineapple is evaporation from the soil. The $K_{c\ mid} < K_{c\ ini}$ since $K_{c\ mid}$ occurs during full ground cover so that soil evaporation is less. Values assume that 50% of the ground surface is covered by black plastic mulch and that irrigation is by sprinkler. For drip irrigation beneath the plastic mulch, K_c's given can be reduced by 0.10.
[24]Includes the water requirements of the shade trees.
[25]The values in this row are similar to those by Johnson et al. (2005).
[26]These $K_{c\ mid}$ and $K_{c\ end}$ values include an implicit K_s (stress) factor of about 0.7 (see Equations 2.43 and 2.45), which is common for wine production. In practice, a K_s-model and estimate should be applied where K_s can range from 0.5 to 1.0. Under no stress, the $K_{c\ mid}$ and $K_{c\ end}$ for wine grapes may equal that for table grapes, depending on plant density, age, and pruning structure.
[27]These $K_{c\ end}$ values represent K_c prior to leaf drop. After leaf drop, $K_{c\ end} \approx 0.20$ for bare, dry soil or dead ground cover and $K_{c\ end} \approx 0.50$ to 0.80 for actively growing ground cover.
[28]For pears having $f_{ceff} = 0.5$, Girona et al. (2004) measured $K_{cb\ mid} = 0.85$, which is estimated using Equations 2.67, 2.72 and 2.73 with $K_{cb\ mid} = 1.1$ and $M_L = 1.5$.
[29]Stone fruit category applies to peaches, apricots, pears, plums and pecans.
[30]The values in this row are derived from Girona et al. (2005) and Ayars et al. (2003) with $f_{ceff} = 0.7$ and and $M_L = 1.5$.
[31]The values in this row are similar to those by Paço et al. (2006) and Ayars et al. (2003) with $f_{ceff} = 0.25$ and $M_L = 1.5$.
[32]The values for citrus are about 20% higher than those reported in FAO-56.
[33]For non-active or only moderately active ground cover (active indicates green and growing ground cover with LAI > about 2), K_c should be weighted between K_c for no ground cover and K_c for active ground cover, with the weighting based on the "greenness" and approximate leaf area of the ground cover.
[34]The values in this row are similar to those by Rogers et al. (1983) for citrus in Florida having Bahia grass cover.

Table 2.5 (continued)

[35]Conifers exhibit substantial stomatal control due to soil water deficit. The K_c can easily reduce below the values presented, which represent well-watered conditions for large forests. (The values in this row are derived from de Azevedo et al. (2003).
[36]Pastor and Orgaz (1994) found monthly K_c for olive orchards having f_c ~ 60% similar to the values shown, except that $K_{c\ mid}$ = 0.45, with stage lengths = 30, 90, 60 and 90 days, respectively for initial, development, midseason and late season periods (see Table 2.6), and using K_c during the winter ("off season") in December to February = 0.50.
[37] The values in this row are similar to those by Villalobos et al. (2000) when $f_{c\ eff}$ of ~ 0.3 to 0.4 are applied.
[38]The values in this row are derived from Testi et al. (2004).[39] These K_c's are for deep water in temperate latitudes where large temperature changes in the water body occur during the year, and initial and peak period evaporation is low as radiation energy is absorbed into the deep water body. During fall and winter periods ($K_{c\ end}$), heat is released from the water body that increases the evaporation above that for grass. Therefore, $K_{c\ mid}$ corresponds to the period when the water body is gaining thermal energy and $K_{c\ end}$ when releasing thermal energy. The higher values for $K_{c\ end}$ represent climates having freezing winter conditions and where ET_0 is low and therefore $K_{c\ end}$ is high These K_c's should be used with caution.

surface covered by vegetation at the time of effective full cover is about 0.7 to 0.8 (Neale et al. 1989; Grattan et al. 1998). Effective full cover for many crops begins at the initiation of flowering. It is understood that the crop or plant can continue to grow in both height and leaf area after the attainment of effective full cover.

The lengths of *initial and development periods* are variable for deciduous trees and shrubs that develop new leaves in the spring. The $K_{c\ ini}$ selected for trees and shrubs should reflect the ground condition prior to leaf emergence or initiation, since $K_{c\ ini}$ is affected by the amount of grass or weed cover, soil wetness, tree density, and mulch density. For example, the $K_{c\ ini}$ for a deciduous orchard having grass ground cover may be as high as 0.8 to 0.9 prior to and during leaf initiation in frost-free climates, whereas the $K_{c\ ini}$ for a deciduous orchard having a bare soil surface may be as low as 0.3 to 0.4 prior to leaf initiation if there is infrequent wetting of the soil by irrigation or by precipitation.

The start of maturity and beginning of decline in K_c starts with senescence at the end of the midseason period when actual ET (ET_a) is reduced relative to ET_0. It is difficult to visualize when this occurs, so the only sure way to know is to measure the ET_a and observe when the K_c begins to decline. Calculations for K_c and ET_a are sometimes presumed to end when the crop is harvested, dries out naturally, reaches full senescence, or experiences leaf drop. For some perennial vegetation in frost free climates, crops may grow year round so that the date of termination is the same as the date of "planting". The length of the late season period may be relatively short (less than 10 days) for vegetation killed by frost (for example maize at high elevations in latitudes >40° N) or for agricultural crops that are harvested fresh (for example table beets and fresh-market

Table 2.6 Parameters used in Equations 2.69–2.71 for estimating $K_{cb\ ini}$, $K_{cb\ mid}$ and $K_{cb\ end}$ in Table 2.5 and using $f_{c\ eff}$ from Table 2.5

Category	$K_{cb\ full}$ - initial	$K_{cb\ full}$ – mid	$K_{cb\ full}$ – end	$K_{c\ min}$	$K_{cb\ cover}$ – initial	$K_{cb\ cover}$ – mid, end	Add[1] to $K_{cb\ ini}$ for $K_{c\ ini}$	Add[1] to $K_{cb\ mid}$ for $K_{c\ mid}$	Add[1] to $K_{cb\ end}$ for $K_{c\ end}$
Almonds	0.20	1.00	0.70[2]	0.15	–	–			
– no ground cover							0.20	0.05	0.05
– ground cover	0.20	1.00	0.70[2]	0.15	0.75	0.80	0.10	0.05	0.05
Apples, cherries, pears	0.30	1.15	0.80[2]	0.15	0.40	0.80			
– killing frost							0.20	0.05	0.05
– no killing frost	0.30	1.15	0.80[2]	0.15	0.75	0.80	0.10	0.05	0.05
Apricots, peaches, pears, plums, pecans	0.30	1.20	0.80[2]	0.15	0.40	0.80			
– killing frost							0.20	0.05	0.05
– no killing frost	0.30	1.20	0.80[2]	0.15	0.70	0.80	0.10	0.05	0.05
Avocado	0.30	1.00	0.90	0.15	–	–			
– no ground cover							0.20	0.05	0.05
– ground cover	0.30	1.00	0.90	0.15	0.75	0.80	0.10	0.05	0.05
Citrus	0.90	0.90	0.90	0.15	0.75	0.80	0.10	0.05	0.05
Mango – no ground cover	0.25	0.85	0.70	0.15			0.10	0.05	0.05
Olives	0.60	0.70	0.60	0.15	0.70	0.70	0.10	0.05	0.05
Pistachios	0.30	1.00	0.70	0.15	0.70	0.70	0.10	0.05	0.05
Walnut	0.40	1.10		0.15	0.75	0.80	0.10	0.05	0.05
Palms – no ground cover	0.85	0.90	0.90	0.15			0.10	0.10	0.10
– ground cover	0.85	0.90	0.90	0.15	0.70	0.70	0.10	0.05	0.05
Grapes – Table or Raison	0.20	1.15	0.90[2]	0.15			0.10	0.05	0.05
Grapes – Wine	0.20	0.80	0.60	0.15			0.10	0.05	0.05

[1]The last three columns are values added to K_{cb} values estimated for the initial, midseason and late season periods during application of Equations 2.69–2.71.

[2]These $K_{c\ full}$ values for end of season represent K_c for full cover conditions prior to leaf drop. After leaf drop, $K_{c\ end} \approx 0.20$ for bare, dry soil or dead ground cover and $K_{c\ end} \approx 0.50$ to 0.80 for actively growing ground cover.

Table 2.7 Mean plant height, h, used in Equations 2.75–2.77 for estimating $K_{cb\ ini}$, $K_{cb\ mid}$ and $K_{cb\ end}$ in Table 2.5 and parameter M_L used in Equation 2.77

	$f_{c\ eff}$					
Category	0.05–0.1	0.25	0.5	0.7	0.9	M_L
Almonds		3	4	5		1.5
Apples, Cherries, Pears		3	3	4		2.0
Apricots, Peaches, Stone Fruit		2.5	3	3	3	1.5
Avocado		3	3	4		2.0
Citrus		2	2.5	3		1.5
Mango		4	4	5		1.5
Olives	2	3	4	4		1.5
Pistachios		2	2.5	3		1.5
Walnut		4	4	5		1.5
Palms	8	8	8	8		1.5
Grapes – Table or Raison		2	2	2		1.5
Grapes – Wine		1.5	1.5	1.5		1.5

vegetables). The value for $K_{c\ end}$ should reflect the condition of the soil surface (average water content and any mulch cover) and the condition of the vegetation following harvest or after full senescence. The K_c during non-growing periods having little or no green ground cover can be estimated using the equation for $K_{c\ ini}$ as described later.

FAO-56 (Allen et al. 1998) provides general lengths of growth (development) stages for a wide variety of crops under different climates and locations. This information is reproduced in Table 2.8. The lengths in Table 2.8 serve only to indicate typical proportions of growing season lengths under a variety of climates. In all applications, local observations of the specific plant stage development should be made to account for local effects of plant variety, climate and cultural practices. Local information can be obtained by interviewing farmers, ranchers, agricultural extension agents and local researchers, by conducting local surveys, or by remote sensing (Heilman et al. 1982; Neale et al. 1989; Choudhury 1994; Tasumi et al. 2005).

2.6.4 Single K_c for the Initial Stage (Annual Crops)

ET during the initial stage for annual crops is predominantly in the form of evaporation from the soil. Accurate estimates for the time-averaged $K_{c\ ini}$ for the mean K_c must consider the frequency that the soil surface is wetted. $K_{c\ mid}$ and $K_{c\ end}$ are less affected by wetting frequency since vegetation during these periods is generally near full ground cover so that effects of surface evaporation are smaller. Figure 2.10a–c from FAO-56 estimate $K_{c\ ini}$ as a function of ET_0, soil type, and wetting frequency

Table 2.8 Lengths of crop development stages* for various planting periods and climatic regions (days). Values are primarily from FAO-56 (Allen et al. 1998) with modification to period lengths for some tree crops

Crop	Init. (L_{ini})	Dev. (L_{dev})	Mid (L_{mid})	Late (L_{late})	Total	Plant date	Region
a. Small Vegetables							
Broccoli	35	45	40	15	135	Sept	Calif. Desert, USA
Cabbage	40	60	50	15	165	Sept	Calif. Desert, USA
Carrots	20	30	50/30	20	100	Oct/Jan	Arid climate
	30	40	60	20	150	Feb/Mar	Mediterranean
	30	50	90	30	200	Oct	Calif. Desert, USA
Cauliflower	35	50	40	15	140	Sept	Calif. Desert, USA
Celery	25	40	95	20	180	Oct	(Semi)Arid
	25	40	45	15	125	Apr	Mediterranean
	30	55	105	20	210	Jan	(Semi)Arid
Crucifers[1]	20	30	20	10	80	Apr	Mediterranean
	25	35	25	10	95	Feb	Mediterranean
	30	35	90	40	195	Oct/Nov	Mediterranean
Lettuce	20	30	15	10	75	Apr	Mediterranean
	30	40	25	10	105	Nov/Jan	Mediterranean
	25	35	30	10	100	Oct/Nov	Arid Region
	35	50	45	10	140	Feb	Mediterranean
Onion (dry)	15	25	70	40	150	Apr	Mediterranean
	20	35	110	45	210	Oct; Jan	Arid Region; Calif.
Onion (green)	25	30	10	5	70	Apr/May	Mediterranean
	20	45	20	10	95	Oct	Arid Region
	30	55	55	40	180	Mar	Calif., USA
Onion (seed)	20	45	165	45	275	Sept	Calif. Desert, USA
Spinach	20	20	15/25	5	60/70	Apr; Sep/Oct	Mediterranean
	20	30	40	10	100	Nov	Arid Region
Radish	5	10	15	5	35	Mar/Apr	Medit.; Europe
	10	10	15	5	40	Winter	Arid Region
b. Vegetables – Solanum Family (Solanaceae)							
Egg plant	30	40	40	20	130\	Oct	Arid Region
	30	45	40	25	140	May/June	Mediterranean
Sweet peppers (bell)	25/30	35	40	20	125	Apr/June	Europe and Medit.
	30	40	110	30	210	Oct	Arid Region
Tomato	30	40	40	25	135	Jan	Arid Region
	35	40	50	30	155	Apr/May	Calif., USA
	25	40	60	30	155	Jan	Calif. Desert, USA
	35	45	70	30	180	Oct/Nov	Arid Region
	30	40	45	30	145	Apr/May	Mediterranean
c. Vegetables – Cucumber Family (Cucurbitaceae)							
Cantaloupe	30	45	35	10	120	Jan	Calif., USA
	10	60	25	25	120	Aug	Calif., USA
Cucumber	20	30	40	15	105	June/Aug	Arid Region
	25	35	50	20	130	Nov; Feb	Arid Region

(continued)

Table 2.8 (continued)

Crop	Init. (L_{ini})	Dev. (L_{dev})	Mid (L_{mid})	Late (L_{late})	Total	Plant date	Region
Pumpkin,	20	30	30	20	100	Mar, Aug	Mediterranean
Winter squash	25	35	35	25	120	June	Europe
Squash, Zucchini	25	35	25	15	100	Apr; Dec	Medit.; Arid
	20	30	25	15	90	May/June	Reg. Medit.; Europe
Sweet melons	25	35	40	20	120	May	Mediterranean
	30	30	50	30	140	Mar	Calif., USA
	15	40	65	15	135	Aug	Calif. Desert, USA
	30	45	65	20	160	Dec/Jan	Arid Region
Water melons	20	30	30	30	110	Apr	Italy
	10	20	20	30	80	Mar/Aug	Near East (desert)
d. Roots and Tubers							
Beets, table	15	25	20	10	70	Apr/May	Mediterranean
	25	30	25	10	90	Feb/Mar	Mediterranean & Arid
Cassava: year 1	20	40	90	60	210	Rainy season	Tropical regions
year 2	150	40	110	60	360		
Potato	25	30	30/45	30	115/130	Jan/Nov	(Semi)Arid
	25	30	45	30	130	May	Climate
	30	35	50	30	145	Apr	Continental
	45	30	70	20	165	Apr/May	Climate
	30	35	50	25	140	Dec	Europe Idaho, USA Calif. Desert, USA
Sweet potato	20	30	60	40	150	Apr	Mediterranean
	15	30	50	30	125	Rainy seas.	Tropical regions
Sugar beet	30	45	90	15	180	Mar	Calif., USA
	25	30	90	10	155	June	Calif., USA
	25	65	100	65	255	Sept	Calif. Desert, USA
	50	40	50	40	180	Apr	Idaho, USA
	25	35	50	50	160	May	Mediterranean
	45	75	80	30	230	Nov	Mediterranean
	35	60	70	40	205	Nov	Arid Regions
e. Legumes (Leguminosae)							
Beans (green)	20	30	30	10	90	Feb/Mar	Calif.
	15	25	25	10	75	Aug/Sept	Mediterranean Calif., Egypt, Lebanon
Beans (dry)	20	30	40	20	110	May/June	Continental
	15	25	35	20	95	June	Climates
	25	25	30	20	100	June	Pakistan, Calif. Idaho, USA
Faba bean,	15	25	35	15	90	May	Europe
broad bean	20	30	35	15	100	Mar/Apr	Mediterranean
dry bean	90	45	40	60	235	Nov	Europe

(continued)

Table 2.8 (continued)

Crop	Init. (L_{ini})	Dev. (L_{dev})	Mid (L_{mid})	Late (L_{late})	Total	Plant date	Region
green bean	90	45	40	0	175	Nov	Europe
Green gram,	20	30	30	20	110	Mar	Mediterranean
cowpeas	25	35	45	25	130	Dry season	West Africa
groundnut							
	35	35	35	35	140	May	High Latitudes
	35	45	35	25	140	May/June	Mediterranean
Lentil	20	30	60	40	150	Apr	Europe
	25	35	70	40	170	Oct/Nov	Arid Region
Peas	15	25	35	15	90	May	Europe
	20	30	35	15	100	Mar/Apr	Mediterranean
	35	25	30	20	110	Apr	Idaho, USA
Soybeans	15	15	40	15	85	Dec	Tropics
	20	30/35	60	25	140	May	Central USA
	20	25	75	30	150	June	Japan
f. Perennial Vegetables (with winter dormancy and initially bare or mulched soil)							
Artichoke	40	40	250	30	360	Apr (1st year)	California
	20	25	250	30	325	May (2nd year)	(cut in May)
Asparagus	50	30	100	50	230	Feb	Warm Winter
	90	30	200	45	365	Feb	Mediterranean
g. Fiber Crops							
Cotton	30	50	60	55	195	Mar–May	Egypt; Pakistan;
	45	90	45	45	225	Mar	Calif.
	30	50	60	55	195	Sept	Calif. Desert,
	30	50	55	45	180	Apr	USA
							Yemen
							Texas
Flax	25	35	50	40	150	Apr	Europe
	30	40	100	50	220	Oct	Arizona
h. Oil Crops							
Castor beans	25	40	65	50	180	Mar	(Semi)Arid
	20	40	50	25	135	Nov	Climates
							Indonesia
Safflower	20	35	45	25	125	Apr	California, USA
	25	35	55	30	145	Mar	High Latitudes
	35	55	60	40	190	Oct/Nov	Arid Region
Sesame	20	30	40	20	100	June	China
Sunflower	25	35	45	25	130	Apr/May	Medit.; California
i. Cereals							
Barley/Oats/ Wheat	15	25	50	30	120	Nov	Central India
	20	25	60	30	135	Mar/Apr	35–45° Latitude
	15	30	65	40	150	July	East Africa

(continued)

Table 2.8 (continued)

Crop	Init. (L_{ini})	Dev. (L_{dev})	Mid (L_{mid})	Late (L_{late})	Total	Plant date	Region
	40	30	40	20	130	Apr	
	40	60	60	40	200	Nov	
	20	50	60	30	160	Dec	Calif. Desert, USA
Winter Wheat	20[2]	60[2]	70	30	180	Dec	Calif., USA
	30	140	40	30	240	Nov	Mediterranean
	160	75	75	25	335	Oct	Idaho, USA
Grains (small)	20	30	60	40	150	Apr	Mediterranean
	25	35	65	40	165	Oct/Nov	Pakistan; Arid Reg.
Maize (grain)	30	50	60	40	180	Apr	East Africa
	25	40	45	30	140	Dec/Jan	Arid Climate
	20	35	40	30	125	June	Nigeria (humid)
	20	35	40	30	125	Oct	India (dry, cool)
	30	40	50	30	150	Apr	Spain (spr, sume-dium); Calif.
	30	40	50	50	170	Apr	
							Idaho, USA
Maize (sweet)	20	20	30	10	80	Mar	Philippines
	20	25	25	10	80	May/June	Mediterranean
	20	30	50/30	10	90	Oct/Dec	Arid Climate
	30	30	30	10[3]	110	Apr	Idaho, USA
	20	40	70	10	140	Jan	Calif. Desert, USA
Millet	15	25	40	25	105	June	Pakistan
	20	30	55	35	140	Apr	Central USA
Sorghum	20	35	40	30	130	May/June	USA, Pakis.
	20	35	45	30	140	Mar/Apr	Medium
							Arid Region
Rice	30	30	60	30	150	Dec; May	Tropics
	30	30	80	40	180	May	Mediterranean
							Tropics
j. Forages							
Alfalfa, total season[4]	10	30	var.	var.	var.		
Alfalfa[4]	10	20	20	10	60	Jan	Calif., USA
1st cutting cycle	10	30	25	10	75	Apr (last −4°C)	Idaho, USA
Alfalfa,[4] other	5	10	10	5	30	Mar	Calif., USA
cutting cycles	5	20	10	10	45	Jun	Idaho, USA
Bermuda for seed	10	25	35	35	105	Mar	Calif. Desert, USA
Bermuda for hay (several cuttings)	10	15	75	35	135	–	Calif. Desert, USA
Grass Pasture[4]	10	20	–	–	–		
Sudan, 1st cutting cycle	25	25	15	10	75	Apr	Calif. Desert, USA

(continued)

Table 2.8 (continued)

Crop	Init. (L_{ini})	Dev. (L_{dev})	Mid (L_{mid})	Late (L_{late})	Total	Plant date	Region
Sudan, other cutting cycles	3	15	12	7	37	June	Calif. Desert, USA
Switchgrass[5]	20	45	40	60	165	Apr	Kansas
k. Sugar Cane							
Sugarcane,	35	60	190	120	405		Low Latitudes
Virgin	50	70	220	140	480		Tropics
	75	105	330	210	720		Hawaii, USA
Sugarcane,	25	70	135	50	280		Low Latitudes
Ratoon	30	50	180	60	320		Tropics
	35	105	210	70	420		Hawaii, USA
l. Tropical Fruits and Trees							
Banana, 1st year	120	90	120	60	390	Mar	Mediterranean
Banana, 2nd year	120	60	180	5	365	Feb	Mediterranean
Pineapple	60	120	600	10	790		Hawaii, USA
m. Grapes and Berries							
Grapes	20	100	90	30	240	Apr	Low
	20	100	90	30	240	Mar	Latitudes
	20	90	50	20	180	May	Calif., USA
	20	90	80	20	210	Apr	High Latitudes
							Mid Latitudes (wine)
Hops	25	40	80	10	155	Apr	Idaho, USA
n. Fruit Trees							
Citrus	90	30	150	95	365	Jan	Mediterranean
Deciduous Orchard	10	10	160	30	210	Mar	High
– light pruning	10	10	190	60	270	Mar	Latitudes
	10	10	190	30	240	Mar	Low Latitudes Calif., USA
Deciduous	10	80	90	30	210	Mar	High
Orchard	10	80	120	60	270	Mar	
– heavy pruning	10	60	140	30	240	Mar	Low Latitudes Calif., USA
Mango	20	40	50	50	160	July	Brazil
Olives	10	20	150	90	270[6]	Mar	Mediterranean
Pistachios	10	20	80	40	150	Feb	Mediterranean
Walnuts	10	10	140	30	190	Apr	Utah, USA
o. Wetlands – Temperate Climate							
Wetlands	10	30	80	20	140	May	Utah, USA;
(Cattails, Bulrush)	180	60	90	35	365	Nov	Killing frost
							Florida, USA
Wetlands (short vegetation)	180	60	90	35	365	Nov	Frost-free climate

(continued)

Table 2.8 (continued)

*Lengths of crop development stages provided in this table are indicative of general conditions, but may vary substantially from region to region, with climate and cropping conditions, and with crop variety. The user is strongly encouraged to obtain appropriate local information.
[1] Crucifers include cabbage, cauliflower, broccoli, and Brussels sprouts. The wide range in lengths of seasons is due to varietal and species differences.
[2] These periods for winter wheat will lengthen in frozen climates according to days having zero growth potential and wheat dormancy. Under general conditions and in the absence of local data, fall planting of winter wheat can be presumed to occur in northern temperate climates when the 30-day running average of mean daily air temperature decreases to 11°C or December 1, whichever comes first. Allen and Robison (2007a, b) reduced canopy development of winter wheat (and amount of K_c above $K_{c\ ini}$) whenever T_{min} was < –25°C and no snow cover was present. They further reduced development by a lesser amount whenever T_{min} was < –10°C as a retardation penalty after a cold freeze. Planting of spring wheat can be assumed to occur when the 30-day running average of mean daily air temperature ending on the planting date increases to 4°C. Spring planting of maize-grain can be assumed to occur when the 30-day running average of mean daily air temperature increases to 10°C.
[3] The late season for sweet maize will be about 35 days if the grain is allowed to mature and dry.
[4] In climates having killing frosts, growing seasons can be estimated using specific temperature or cumulative growing degree days or running average air temperature to begin and killing frosts to terminate. For example for alfalfa and grass:
alfalfa: last –4°C in spring until first –4°C in fall (Everson et al. 1978) or accumulation of 240 degree-days since January 1 using 0° base for green up and first occurrence of –7°C frost in fall (Allen and Robison 2007a).
grass: 7 days before last –4°C in spring and 7 days after last –4°C in fall (Kruse and Haise 1974) or 30 day running average mean air temperature for period ending on day of green up = 4°C and first occurrence of –5°C frost in fall (Allen and Robison 2007a)
[5] Based on measurements of ET from prairie in Kansas by Verma et al. (1991) comprised of switchgrass (*Panicum virgatum*), big bluestem (*Andropogon gerardii*) and indiangrass (*Sorghastrum nutans*).
[6] Olive trees gain new leaves in March and often have transpiration during winter, where the K_c continues outside of the "growing period" and total season length may be set to 365 days.

and represent an improvement relative to those early presented earlier by Doorenbos and Pruitt (1977). Equations for these curves are given in Allen et al. (1998, 2005b). These authors and Pereira and Alves (2005) present a numerical computation procedure for $K_{c\ ini}$ useful for model applications. Figure 2.10a is used for all soil types when wetting events (precipitation and irrigation) are light (i.e., infiltrated depths average about 10 mm per wetting event), Fig. 2.10b is used for "heavy" wetting events, where infiltrated depths are greater than 30–40 mm, on coarse-textured soils,[2] and Fig. 2.10c is used for heavy wetting events on fine and medium-textured soils. In general, the mean time interval is estimated by counting all rainfall and irrigation events occurring during the initial period that are greater than the medium. Wetting events occurring on adjacent days are typically counted as one event. When average

[2] Coarse-textured soils include sands and loamy sand textured soils. Medium-textured soils include sandy loam, loam, silt loam and silt textured soils. Fine-textured soils include silty clay loam, silty clay and clay textured soils.

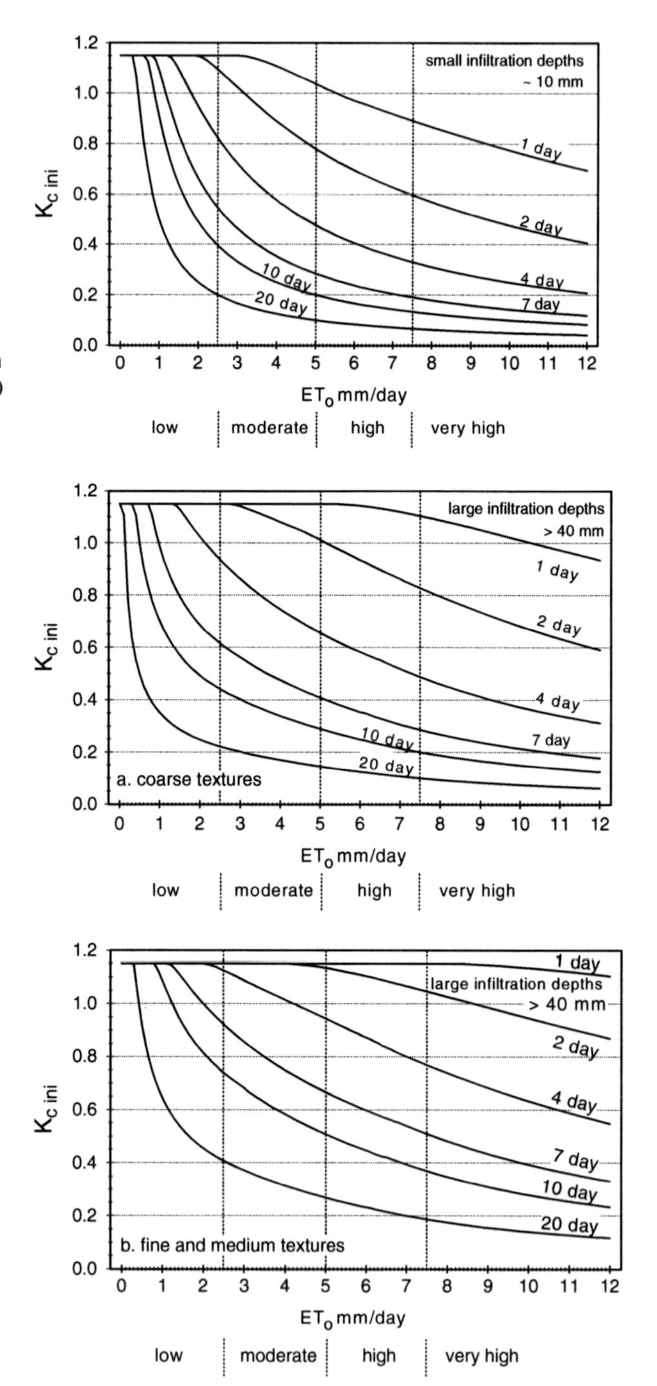

Fig. 2.10 Average $K_{c\ ini}$ for the initial crop development stage as related to the level of ET_0 and the interval between irrigations and/or significant rain during the initial period for (top) all soil types when wetting events are light (about 10 mm per event); (middle) coarse-textured soils when wetting events are greater than about 40 mm; and (bottom) medium and fine-textured soils when wetting events are greater than about 40 mm (From Allen et al. 1998, 2005b, c)

infiltration depths are between 10 and 40 mm, the value for $K_{c\ ini}$ can be interpolated between Fig. 2.10a, b or Fig. 2.10c.

The $K_{c\ ini}$ values in Fig. 2.10 should be corrected for the fraction of ground from where evaporation occurs, and the depth of infiltrated water should also be corrected by dividing by this same fraction. The fraction corresponds to the soil wetted fraction, f_w, that varies between 1.0 for rain, sprinkler and basin irrigation to 0.3–0.7 for drip irrigation (Table 2.10). The total infiltrated water, I_w (mm), is presumed to infiltrate within the fraction f_w of the surface, thus the total infiltrated water to be used and interpolated with Fig. 2.10a–c should be I_w/f_w. The $K_{c\ ini}$ value obtained from the figures should then be corrected as $f_w\ K_{c\ ini}$. The value for f_w should also be corrected using the concept of the f_{ew} fraction that is described in a following section on the dual K_c procedure.

2.6.5 K_c Curves for Forage Crops

Many crops grown for forage or hay receive multiple harvests during the growing season. Each harvest essentially terminates a "sub" growing season and associated K_c curve and initiates a new "sub" growing season and associated K_c curve. The resulting K_c curve for the entire growing season is the aggregation of a series of K_c curves associated with each sub cycle. A K_c curve constructed for alfalfa grown for hay in southern Idaho is illustrated in Fig. 2.11. Cuttings may create a ground surface with less than 10% vegetation cover. Cutting cycle 1 may have longer duration

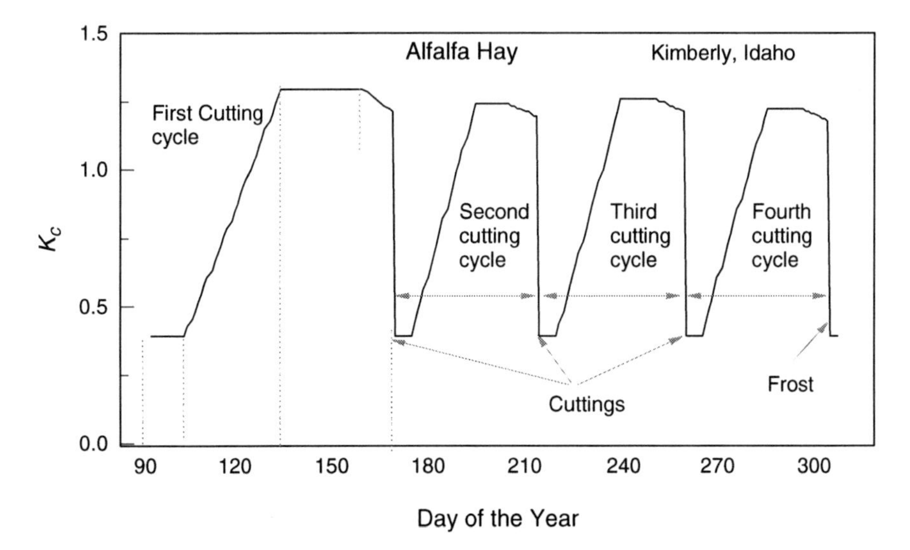

Fig. 2.11 Crop coefficient curve for alfalfa hay crop in southern Idaho having four cuttings

than cycles 2, 3 and 4 if low air and soil temperatures or short day length during this period moderate the crop growth rate. Frosts terminate the growing season in southern Idaho sometime in the fall, usually around day of year 280 to 300 (early to mid October, see footnote 4 of Table 2.8). Magnitudes of K_c values during mid-cycle periods for each cutting cycle change as a result of adjusting $K_{c\,mid}$ and $K_{c\,end}$ for each period using Equation 2.57. Basal K_{cb} curves for forage or hay crops can be constructed similar to Fig. 2.11.

2.7 The Dual K_c Method: Incorporating Specific Wet Soil Effects

2.7.1 *Theory*

The previous sections use a 'single' K_c where any time-averaged effects of evaporation from the soil surface are averaged into the K_c value. The single or 'mean' K_c represents, on any particular day, average evaporation rates from the soil and plant surfaces. An alternative K_c approach is the 'dual' K_c method, where the K_c value is divided into a 'basal' crop coefficient, K_{cb}, and a separate component, K_e, representing evaporation from the soil surface. The basal crop coefficient represents *actual ET* conditions when the soil surface is dry, but sufficient root zone moisture is present to support full transpiration. Generally, a daily calculation time-step is required to apply the dual K_c method, whereas the mean K_c method can be applied on daily, weekly or monthly time steps.

The mean K_c approach is used for planning studies and irrigation systems design where averaged effects of soil wetting are appropriate. For typical irrigation management, this use is valid. The dual K_c approach, which requires more numerical calculations, is suited for research studies where specific effects of day to day variation in soil surface wetness and the resulting impacts on daily *actual* ET_{act}, the soil water profile, and deep percolation fluxes are important. The form of the equation for $K_{c\,act}$ in the dual K_c approach is:

$$K_{c\,act} = K_{cb} + K_e \qquad 2.58$$

where K_{cb} is the basal crop coefficient [0 to 1.4], and K_e is a soil water evaporation coefficient [0 to 1.4]. Both terms are dimensionless. K_{cb} is defined as the ratio of ET_p to ET_{ref} when the soil surface layer is dry, but where the average soil water content of the root zone is adequate to sustain full plant transpiration.

The dual K_c method introduced in this section is based on procedures developed in FAO-56. The K_e component of the dual K_c method describes the evaporation component of ET_p. Because the dual K_c method incorporates the effects of specific wetting patterns and frequencies that may be unique to a single field, this method can provide more accurate estimates of evaporation components and total ET on an individual field basis.

When the soil surface layer is wet, following rain or irrigation, K_e is at a maximum, and when the soil surface layer is dry, K_e is small and can approach zero. When the soil is wet, evaporation occurs at some maximum rate where $K_{cb} + K_e$ is limited by a maximum value $K_{c,max}$:

$$K_e = K_r\,(K_{c\ \max} - K_{cb}) \le f_{ew}\,K_{c\ \max} \qquad 2.59$$

where $K_{c\ max}$ is the maximum value of K_c following rain or irrigation, K_r is a dimensionless evaporation reduction coefficient [defined later] and is dependent on the cumulative depth of water depleted (evaporated), and f_{ew} is the fraction of the soil that is both exposed to solar radiation and that is wetted. The evaporation rate is restricted by the estimated amount of energy available at the exposed soil fraction, i.e., K_e cannot exceed $f_{ew}\ K_{c\ max}$. $K_{c\ max}$ for the ET_0 basis ($K_{c\ max}$) ranges from about 1.05 to 1.30 (Allen et al. 1998, 2005b):

$$K_{c\ \max} = \max\left(\begin{matrix}\left\{1.2 + \left[0.04\,(u_2 - 2) - 0.004\,(RH_{\min} - 45)\right]\left(\frac{h}{3}\right)^{0.3}\right\}, \\ \left\{K_{cb} + 0.05\right\}\end{matrix}\right) \qquad 2.60$$

where h (m) is the mean plant height during the growth period (initial, development, mid-season, or late-season), and the 'max ( )' function indicates the selection of the maximum of values separated by the comma. Equation 2.60 ensures that $K_{c\ max}$ is always greater than or equal to the sum K_{cb} + 0.05, suggesting that wet soil increases the K_c value above K_{cb} by 0.05 following complete wetting of the soil surface, even during periods of full ground cover.

The surface soil layer can dessicate to an air dry water content approximated as half of the difference between the wilting point, θ_{WP}, and oven dry. The amount of water that can be removed by evaporation during a complete drying cycle is estimated as:

$$TEW = 1000\,(\theta_{FC} - 0.5\theta_{WP})\,Z_e \qquad 2.61$$

where TEW (total evaporable water) in mm is the maximum depth of water that can be evaporated from the surface soil layer assuming that the soil was completely wetted. Field capacity, θ_{FC}, and θ_{WP} are expressed in $m^3\ m^{-3}$ and Z_e is the effective depth in m of the surface soil subject to drying to 0.5 θ_{WP} by way of evaporation. Typical values for θ_{FC}, θ_{WP}, REW and TEW are given in Table 2.9 for a range in soil types. Z_e is an empirical value based on observation. Some evaporation or soil drying will be observed to occur below the Z_e depth. FAO-56 recommended values for Z_e of 0.10–0.15 m, with 0.1 m recommended for coarse soils and 0.15 m recommended for fine textured soils.

Table 2.9 Typical soil water characteristics for different soil types (From Allen et al. 1998)

Soil type (USDA Soil Texture Classification)	Soil water characteristics			Evaporation parameters		
	θ_{FC}	θ_{WP}	$(\theta_{FC}-\theta_{WP})$	Amount of water that can be depleted by evaporation		
				Stage 1	Stages 1 and 2	Stages 1 and 2
				REW	TEW* (Z_e = 0.10 m)	TEW* (Z_e = 0.15 m)
	$m^3\ m^{-3}$	$m^3\ m^{-3}$	$m^3\ m^{-3}$	mm	mm	mm
Sand	0.07–0.17	0.02–0.07	0.05–0.11	2–7	6–12	9–13
Loamy sand	0.11–0.19	0.03–0.10	0.06–0.12	4–8	9–14	13–21
Sandy loam	0.18–0.28	0.06–0.16	0.11–0.15	6–10	15–20	22–30
Loam	0.20–0.30	0.07–0.17	0.13–0.18	8–10	16–22	24–33
Silt loam	0.22–0.36	0.09–0.21	0.13–0.19	8–11	18–25	27–37
Silt	0.28–0.36	0.12–0.22	0.16–0.20	8–11	22–26	33–39
Silt clay loam	0.30–0.37	0.17–0.24	0.13–0.18	8–11	22–27	33–40
Silty clay	0.30–0.42	0.17–0.29	0.13–0.19	8–12	22–28	33–42
Clay	0.32–0.40	0.20–0.24	0.12–0.20	8–12	22–29	33–43

*$TEW = (\theta_{FC} - 0.5\ \theta_{WP})\ Z_e$.

Table 2.10 Common values for the fraction of soil surface wetted by irrigation or precipitation (After FAO-56; Allen et al. 1998)

Wetting event	f_w
Precipitation	1.0
Sprinkler irrigation, field crops	1.0
Sprinkler irrigation, orchards	0.7–1.0
Basin irrigation	1.0
Border irrigation	1.0
Furrow irrigation (every furrow), narrow bed	0.6–1.0
Furrow irrigation (every furrow), wide bed	0.4–0.6
Furrow irrigation (alternated furrows)	0.3–0.5
Micro spray irrigation, orchards	0.5–0.8
Micro irrigation (trickle and drip)	0.3–0.4

The K_r coefficient of Equation 2.59 is calculated as:

$$K_r = 1.0 \quad \text{for } D_{e,j-1} \le REW \qquad 2.62a$$

$$K_r = \frac{TEW - D_{e,j-1}}{TEW - REW} \text{ for } D_{e,j-1} > REW \qquad 2.62b$$

where $D_{e,j-1}$ is cumulative depletion from the soil surface layer at the end of day j-1 (the previous day) [mm], and TEW and REW are in mm ($REW < TEW$). Evaporation from the soil beneath the crop canopy, occurring at a slower rate, is assumed included in the basal K_{cb} coefficient.

In the FAO-56 (Allen et al. 1998) model, the term f_w is defined as the fraction of the surface wetted by irrigation and/or precipitation. This term defines the potential spatial extent of evaporation. Common values for f_w are listed in Table 2.10 and illustrated in Fig. 2.12. When the soil surface is completely wetted, as by

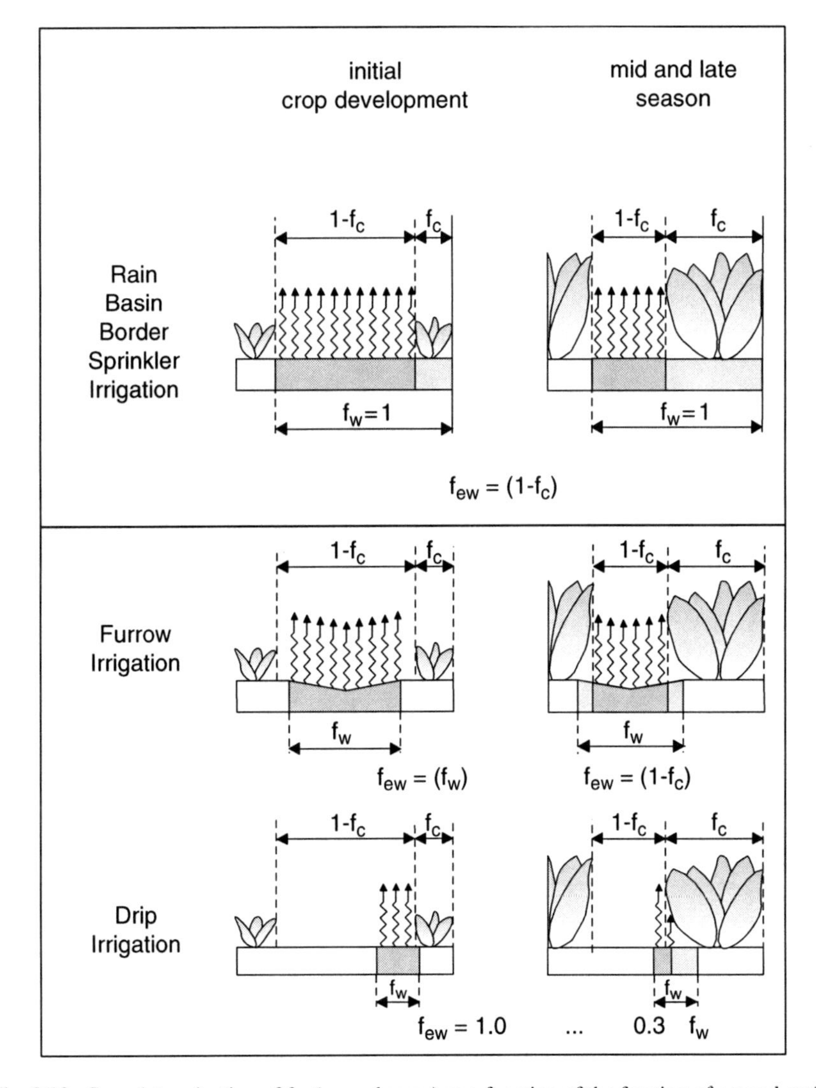

Fig. 2.12 Crop determination of f_{ew} (greyed areas) as a function of the fraction of ground surface coverage (f_c) and the fraction of the surface wetted (f_w) (From Allen et al. 1998)

precipitation or sprinkler, f_{ew} of Equation 2.59 is set equal to $(1 - f_c)$, where f_c is the fraction of soil surface effectively covered by vegetation. For irrigation systems where only a fraction of the ground surface (f_w) is wetted, f_{ew} is limited to f_w:

$$f_{ew} = \min(1 - f_c, f_w) \tag{2.63}$$

Both $(1 - f_c)$ and f_w, for numerical stability, have limits of [0.01–1]. In the case of drip irrigation, Allen et al. (1998) suggest that where the majority of soil wetted by irrigation is beneath the crop canopy and is shaded, f_w be reduced to about one-half to one-third of that given in Table 2.10. Their general recommendation for drip irrigation is to multiply f_w by $[1 - (2/3)f_c]$. Pruitt et al. (1984) and Bonachela et al. (2001) described evaporation patterns and drying extent under drip irrigation.

The value for f_c is limited to <0.99 for numerical stability and is generally determined by visual observation. For purposes of estimating f_{ew}, f_c can be estimated from K_{cb} as:

$$f_c = \left(\frac{K_{cb} - K_{c\,min}}{K_{c\,max} - K_{c\,min}} \right)^{(1+0.5h)} \tag{2.64}$$

where f_c ranges between 0 and 0.99 and $K_{c\,min}$ is the minimum K_c for dry bare soil with no ground cover. The difference $K_{cb} - K_{c\,min}$ is limited to ≥0.01 for numerical stability. The value for f_c will change daily as K_{cb} changes. $K_{c\,min}$ ordinarily has the same value as $K_{cb\,ini}$ used for annual crops under nearly bare soil conditions (i.e., $K_{c\,min} \sim 0.15$). However, $K_{c\,min}$ is set to 0 or nearly zero under conditions with large time periods between wetting events, for example in applications with natural vegetation in deserts. The value for f_c decreases during the late season period in proportion to K_{cb} to account for local transport of sensible heat from senescing leaves to the soil surface.

2.7.2 *Water Balance of the Soil Surface Layer*

Estimation of K_e requires a daily water balance for the f_{ew} fraction of the surface soil layer. The daily soil water balance equation is:

$$D_{e,j} = D_{e,j-1} - (P_j - RO_j) - \frac{I_j}{f_w} + \frac{E_j}{f_{ew}} + T_{ei,j} + DP_{ei,j} \tag{2.65}$$

where $D_{e,j-1}$ and $D_{e,j}$ are cumulative depletion depth in mm at the ends of days j-1 and j, P_j and RO_j are precipitation and precipitation runoff in mm from the soil surface on day j, I_j is the irrigation depth in mm on day j that infiltrates the soil, E_j is evaporation in mm on day j (i.e., $E_j = K_e\ ET_0$), $T_{ei,j}$ is the depth of transpiration in mm from the exposed and wetted fraction of the soil surface layer on day j, and $DP_{ei,j}$ is the deep percolation in mm from the f_{ew} fraction of the soil surface layer on day j if soil water content exceeds field capacity. *RO* can be estimated using the curve number (USDA-SCS 1972, Hawkins et al. 1985) or other means.

Assuming that the surface layer is at field capacity following heavy rain or irrigation, the minimum value for $D_{e,j}$ is zero. The limits imposed on $D_{e,j}$ are consequently $0 \le D_{e,j} \le TEW$. It is recognized that water content of the soil surface layer can exceed *TEW* for short periods of time while drainage is occurring. However, because the length of time that this occurs varies with soil texture, wetting depth, and tillage, $D_{e,j} \ge 0$ is assumed. Additionally, it is recognized that some drainage in soil occurs at very small rates at water contents below field capacity. To some extent, impacts of these simple assumptions can be compensated for, if needed, in setting the value for Z_e or *TEW*. The irrigation depth I_j is divided by f_w to approximate the infiltration depth to the f_w portion of the soil surface. Similarly, E_j is divided by f_{ew} because it is assumed that all E_j (other than residual evaporation implicit to the K_{cb} coefficient) is taken from the f_{ew} fraction of the surface layer.

2.7.2.1 Transpiration and Deep Percolation from the Surface layer

The amount of transpiration extracted from the f_{ew} fraction of the evaporating soil layer is generally a small fraction of total transpiration, and is generally ignored. For shallow-rooted annual crops where the depth of the maximum rooting is less than about 0.5 m, T_e may impact the water balance of the surface layer and should be considered. Instructions on estimating T_e are given by Allen et al., (2005c).

In the simple water balance procedure of FAO-56 (Allen et al. 1998), it is assumed that the soil water content is limited to $\le \theta_{FC}$ on the day of a complete wetting event. This is a reasonable assumption considering the shallowness of the surface layer. Downward drainage (percolation) of water from the surface layer (the top 0.1 to 0.15 m of soil) is calculated as:

$$DP_{e,j} = (P_j - RO_j) + \frac{I_j}{f_w} - D_{e,j-1} \ge 0 \qquad 2.66$$

$DP_{e,j}$ is not to be confused with deep percolation from the root zone. Most of $DP_{e,i}$ may be captured by the underlying root zone.

2.7.2.2 Initialization of the Water Balance and Order of Calculation

To initiate the water balance for the evaporating layer, the user can assume that the soil surface layer is near θ_{FC} following a heavy rain or irrigation so that $D_{e,j-1} = 0$. Where a long period of time has elapsed since the last wetting, the user can assume that all evaporable water has been depleted from the evaporation layer at the beginning of calculations so that $D_{e,j-1} = TEW = 1{,}000\ (\theta_{FC} - 0.5\ \theta_{WP})Z_e$. Calculations for the dual $K_{cb} + K_e$ procedure, for example when using a spreadsheet, proceed in the following order: K_{cb}, h, $K_{c\,max}$, f_c, f_w, f_{ew}, K_r, K_e, E, DP_e, D_e, I, K_c, and ET_0.

2.7.2.3 Conditions for Maximum Transpiration

The K_c values in Table 2.5 represent potential water consumption by healthy, relatively disease free, and densely planted stands of vegetation having adequate levels of soil water. When stand density, height, or leaf area are significantly less than that attained under ideal or normal (pristine) conditions, the value for K_c may be reduced. Low stand density, height and leaf area are caused by disease, low soil fertility, high soil salinity, water logging or water shortage (moisture stress), or by poor stand establishment. The reduction in the value for K_c during the midseason for poor crop stands can be as much as 0.3 to 0.5 for extremely poor crop stands and can be approximated according to the amount of effective (green) leaf area relative to that for healthy vegetation having normal planting densities. Procedures for reducing K_c according to the reduction in leaf area and the fraction of ground cover are given in Chapter 9 of FAO-56 (Allen et al. 1998).

2.7.3 *Application of the Basal K_{cb} Procedure over a Growing Season*

The first step in applying the basal K_{cb} approach is to construct the K_{cb} curve using $K_{cb\ ini}$, $K_{cb\ mid}$ and $K_{cb\ end}$ similar to constructing the single or mean K_c curve. Equations for computing K_e (and Stress coefficient K_s if necessary) are applied on a daily calculation time step where daily K_{cb} is interpolated from the constructed K_{cb} curve. An illustration of applying the $K_{cb} + K_e$ procedure for a snap bean crop harvested for dry seed are shown in Fig. 2.13. The measured ET_p data are from a precision lysimeter system at Kimberly, Idaho (J.L.Wright 1990 personal communication and Van der Kimpen 1991). The soil at Kimberly had a silt loam texture. Soil evaporation parameters were $Z_e = 0.15$ m, TEW = 34 mm and REW = 8 mm. Nearly all wetting events were from alternate row furrow irrigation so that the value for f_w was set to 0.5. Irrigation events occurred at about midday or during early afternoon. The agreement between the estimated values for daily K_c from Equation 2.68 (thin, continuous line) and actual 24-h measurements (symbols) is relatively good.

2.7.3.1 Water Balance of the Root Zone

Often, precipitation and irrigation amounts are not sufficient to supply the full ET_p requirement. In these situations, soil water content and potential energy of water in the root zone reduce to levels too low to permit plant roots to extract the full ET_p amount. Under these conditions, water stress is said to occur, and the ET_a becomes less than the ET_p. The reduction in ET_a can be estimated using a daily soil water

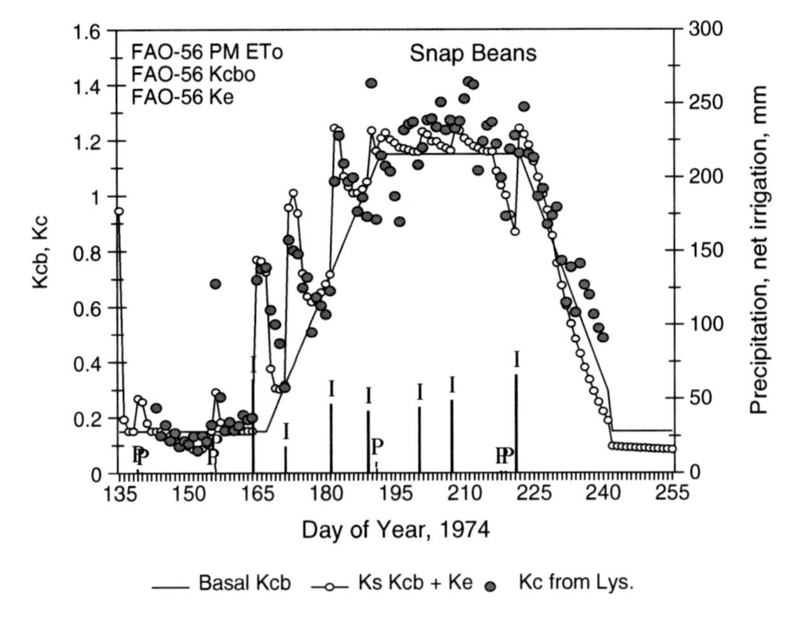

Fig. 2.13 Measured (circles) and estimated daily crop coefficients for a snap bean crop at Kimberly, Idaho. The basal crop curve (K_{cb}) was derived from K_c values given in Table 2.5 (P = precipitation event and, I = irrigation (Data from J. Low Wright 1990))

balance, as follows, or it can be estimated in a more general way as shown in Chapter 4 using depleted fraction of the field or system. When field-specific estimates of ET_a are needed, they can be estimated by reexpressing Eq. 2.54 as:

$$ET_a = K_s K_c ET_0 \tag{2.67}$$

or, if applying the dual K_c approach, reexpressing Eq. 2.58 as:

$$K_{c\,act} = K_s K_{cb} + K_e \tag{2.68}$$

where $ET_a = K_{c\,act}\, ET_o$. The stress coefficient, K_s is estimated following FAO-56 (Allen et al., 1998) as:

$$K_s = \frac{TAW - D_r}{TAW - RAW} = \frac{TAW - D_r}{(1-p)\,TAW} \tag{2.69}$$

for $D_r > RAW$, where D_r is the root zone depletion, defined as water shortage relative to field capacity. *RAW* is the 'readily' available water (mm), *TAW* is total available soil water in the root zone (mm), and p (0 to 1) is the fraction of *TAW* that a crop can extract from the root zone without suffering water stress. *TAW* is estimated as the difference between the water content at field capacity and wilting point:

$$TAW = 1000\ (\theta_{FC} - \theta_{WP})\ Z_r \qquad 2.70$$

where Z_r is the effective rooting depth (m) and Z_r contains Z_e. *RAW* is estimated as:

$$RAW = p\ TAW \qquad 2.71$$

where *RAW* has units of *TAW* (mm). When $D_r \leq RAW$, $K_s = 1$. At field capacity, $D_r = 0$. The degree of stress is presumed to progressively increase as D_r increases past *RAW*, the depth of readily available water in the root zone. The value for *p* ranges from about 0.4 for shallow-rooted crops to 0.6 for deep rooted crops. Recommended values for specific crops can be found in FAO-56 (Allen et al., 1998). Table 4.2 lists values for Z_r for a number of crops and Chapter 4 describes means to estimate the increase in Z_r with time for annual crops.

The calculation of K_s requires a daily water balance computation for the root zone, and is done on a field-by-field basis. A daily water balance, expressed in terms of depletion at the end of the day, is:

$$D_{r,i} = D_{r,i-1} - (P - RO)_i - I_i - q_i + ET_{a,i} + DP_i \qquad 2.72$$

where,

$D_{r,i}$ = root zone depletion at the end of day *i*, mm
$D_{r,\,i-1}$ = depletion in the root zone at the end of the previous day, *i*-1, mm
P_i = precipitation on day *i*, mm
RO_i = runoff from the soil surface on day *i*, mm
I_i = net irrigation depth on day *i* that infiltrates the soil, mm
q_i = capillary rise from the groundwater table on day *i*, mm
$ET_{a,\,i}$ = actual crop evapotranspiration on day *i*, mm
DP_i = water flux out of the root zone by deep percolation on day *i*, mm.

The capillary rise, q_i, is estimated in Chapter 4 and estimation of surface runoff is described in Chapter 3. Although following heavy rain or irrigation, soil water content might temporally exceed field capacity, in equation 2.72, the total amount of water exceeding field capacity is assumed to be lost the same day via deep percolation, following any ET_a for that day. This does permit the extraction of one day's ET_a from this excess before percolation. The root zone depletion will gradually increase as a result of ET_a. In the absence of a wetting event, the root zone depletion will reach the value *TAW* defined from Equation 2.70. At that moment no water is left for ET_a, and K_s and ET_a become zero.

2.7.3.2 Deep Percolation from the Root Zone

Following heavy rain or irrigation, the soil water content in the root zone may exceed field capacity. In applications of Equation 2.72, *DP* is assumed to occur within the same day of a wetting event, so that the depletion $D_{r,\,i}$ in Equation 2.72 becomes zero. Therefore,

$$DP_i = (P_i - RO_i) + I_i - ET_{c,i} - D_{r,i-1} \geq 0 \qquad 2.73$$

As long as the soil water content in the root zone is below field capacity (i.e., $D_{r,i} > 0$), the soil is assumed to not drain and $DP_i = 0$. The DP_i term in Equations 2.72 and 2.73 is not to be confused with the $DP_{e,i}$ term used in Equations 2.66 and 2.67 for the evaporation layer. Both terms can be calculated at the same time, but are independent of one another.

2.8 ET During the Non-growing Season

During non-growing periods, *actual ET* is dominated by evaporation, rather than transpiration, especially if the non-growing season is caused by killing frosts. Non-growing season ET_a is therefore generally best estimated using techniques that accurately estimate evaporation (*E*) from the soil surface.

Evaporation is a strong function of wetting frequency and reference *ET* rate and therefore, the $K_{c\,ini}$ estimated from Fig. 2.10 can be used as an estimate of K_c during the non-growing season (Martin and Gilley 1993), with some adjustment for impacts of surface cover by dead or dormant vegetation, as described later in Section 2.11. Alternatively, the K_c during the non-growing season can be estimated using the dual $K_{cb} + K_e$ method with some adjustment to *TEW* and *REW* during cold periods as described by Allen et al. (2005b, c). The dual procedure was applied by Allen and Robison (2007a, b) at weather locations throughout Idaho for complete calendar years including winter, with adjustments made during periods of snow cover. Snyder and Eching (2004, 2005) suggested a procedure for combining the K_c during the non-growing season, estimated using a procedure similar to Fig. 2.10, with the K_c curve for the growing season to create a continuous K_c for the entire calendar year. This was done by taking the maximum, for each day, of smoothed values for $K_{c\,ini}$ as estimated from Fig. 2.10 or similar curve and the K_c obtained from the growing season $K_{c\,ini}$ curve.

As discussed in Section 2.6, the ET_a can also be quantified with satellite remote sensing. Figure 2.14 shows the actual evapotranspiration for three land uses in the Roxo basin in south Portugal, as quantified through the energy balance of the pixels of Landsat and Modis images. Please note that ET_a is given for the full year and not just for the crop growing season, because water also evaporates when there is no crop in the field.

2.9 Evapotranspiration from Landscapes

Over the past several decades, the water requirements and water consumption by residential and urban landscapes have become increasingly important in terms of quantity and value of water consumption. Procedures similar to those from agriculture have been adapted to estimate *ET* from landscapes. However, two distinctions between agriculture and landscapes exist regarding *ET* quantification:

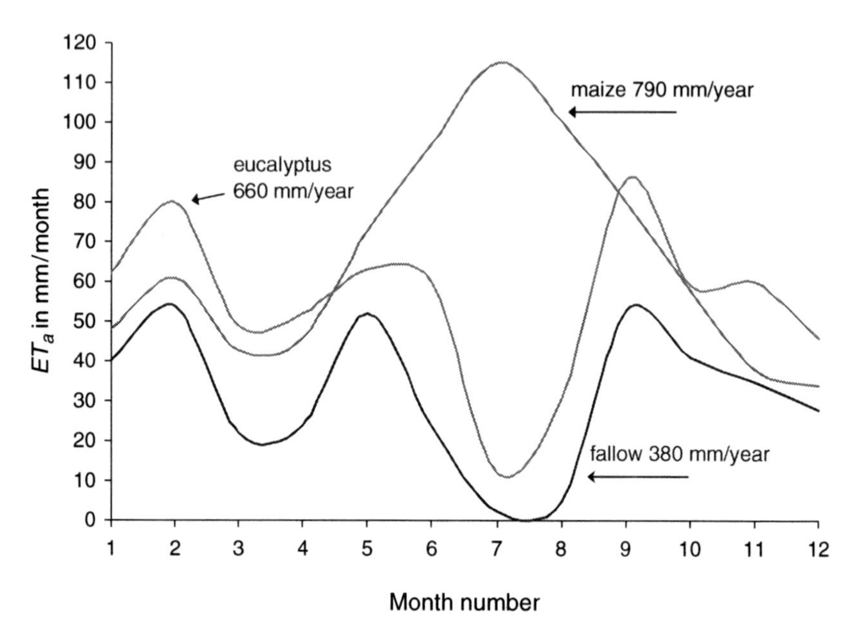

Fig. 2.14 Actual evapotranspiration for three alternative land uses in the Roxo basin (approximately 38° N, 8° W), Portugal, where average precipitation is 535 mm/year

- Landscape systems are nearly always comprised of a mixture of multiple types and species of vegetation, thereby complicating the estimation of ET_a.
- Typically, the objective of landscape irrigation is to promote appearance rather than biomass production, as is the case in agriculture.

Therefore, the target ET_a for landscapes may include an intentional "water stress" where landscape plants are watered less than they would be if they were irrigated like a crop. They are watered enough to look good and to survive, but the plants are stressed and will not be at maximum productivity. This adjustment can result in an ET_a being significantly lower than the potential value. The magnitude of the water stress depends on physiological and morphological requirements of the plants; the goal is to sustain health and appearance with minimal irrigation. For example, water conservation studies on turfgrass have demonstrated water savings of 30% for cool-season turfgrasses and 40% for warm-season turfgrasses may be attainable without significant loss of quality (Pittenger and Shaw 2001). Many shrubs and groundcovers can be managed for even more stress-induced reduction in ET_a. A third departure of landscape ET_a from agricultural ET_p is that few landscape sites meet the "extensive surface" requirement needed to insure the equilibrium between the lower boundary layer of the atmosphere and the vegetation that is implied in the Penman-Monteith equation. Therefore, compensating adjustments are necessary to the landscape coefficient in the form of a microclimate factor to account for effects of local surroundings.

Because of the frequent inclusion of water stress in target *ET* values for landscape design and management, distinction must be made between these target *ET* values and ET_a values. The ET_a values may exceed target *ET* values if the landscape receives more water than required by the target that includes intentional stress. Under these conditions, landscape vegetation may exploit the additional available water, subject to some limit constrained by environmental energy for evaporation and leaf area. This limit, which follows behavior and principles used for agricultural crops, may exceed the targeted *ET* rate for the particular landscape cover. Conversely, ET_a may be less than targeted *ET* values if actual stress levels to the landscape are more excessive than targeted. Therefore, two *ET* values for landscape are distinguished here. The first is the *target landscape ET*, referred to as ET_L, which is based on minimum *ET* levels, relative to climate, necessary to sustain a healthy, attractive landscape. The second *ET* value is the *actual landscape ET*, $ET_{L,a}$, which is based on landscape type and on actual water availability.

The target *ET* for a landscape is calculated as

$$ET_L = K_L ET_0 \qquad 2.74$$

where ET_L is the target landscape *ET* (in mm day^{-1}, mm month^{-1}, or mm year^{-1}), and ET_0 is the grass reference *ET* in the same units. K_L is the target landscape coefficient, similar to the crop coefficient used in agricultural applications.

There has been relatively limited experimental research on quantifying water needs of the diverse array of landscape plant types (Pittenger and Henry 2005). Much of the existing information is based largely on observation rather than on scientifically obtained data. Some of the leading work on landscape *ET* has been done in California, where water applied to landscapes in southern California is estimated to be 25–30% of all water used in the region (Pittenger and Shaw 2001). Pittenger and Shaw (2007), unpublished work from University of California Cooperative Extension) produced a table of K_L values for 35 landscape groundcovers and shrubs that provide acceptable landscape performance after establishment, but that cause a managed amount of moisture stress via limited water application. Costello et al. (2000) and IA (2005) described a recent procedure termed WUCOLS (water use classification of landscape species), where the K_L has been decoupled into reproducible and visually apparent components representing the effects of three or four factors that control the value for K_L. The decoupling was done to provide for application to the wide diversity of vegetation types and environments of landscape systems. Snyder and Eching (2004, 2005) have proposed a similar decoupled procedure for estimating a formulated K_L, but which uses different ranges for the components. The procedure of Snyder and Eching was refined and formally expressed in quantitative formulae by Allen et al. (2007), where K_L was decoupled into multiplicative terms describing the effects of vegetation type, vegetation density, managed stress, and local microclimate.

As stated previously, typically, the objective of landscape irrigation is to promote appearance rather than biomass production, as is the case in agriculture. Therefore, the target ET_a for landscapes may include an intentional and managed "stress" factor in the baseline value for ET_L, where landscape plants are watered less than they would be if they were irrigated like a crop. This management is done by adjusting irrigation water schedules to apply less water than the vegetation will potentially transpire. The magnitude of the stress factor depends on physiological and morphological requirements of the plants. For example, water conservation studies on turfgrasses have demonstrated that water savings of up to 30% for cool-season turfgrasses and 40% for warm-season turfgrasses may be attainable without significant loss of quality (Pittenger and Shaw 2001). Many woody shrubs and groundcovers can be managed for even more stress-induced reduction in ET (Kjelgren et al. 2000).

Pittenger et al. (2001), Shaw and Pittenger (2004) and Pittenger and Shaw (2007), personal communication) defined water needs of non-turf landscape plants as a percentage of ET_0 needed to maintain their appearance and intended function (e.g. shade, green foliage, screening element). In the procedure of Snyder and Eching (2005) and Allen et al. (2007), the landscape coefficient K_L is decoupled into components that describe the impacts of vegetation type, density of vegetation, microclimatic effects and managed stress factor. The managed water stress ratio, ET_a/ET_p, represents the fraction of the potential ET rate targeted to obtain the functional and visual characteristics of the landscape vegetation. The ratio ET_a/ET_p has a range of 0 to 1.0 where 1.0 represents conditions of no moisture stress (and no real water conservation) and 0 represents complete lapse of plant transpiration and probable plant death. High values for ET_a/ET_p will sustain relatively lush, high leaf-area vegetation stands that may be necessary to sustain long-term plant health or appearance. Low values for ET_a/ET_p represent substantial managed plant water stress and reduction in ET_a, generally at the cost of biomass accumulation and potentially visual effects (Richie and Pittenger 2000).

Typically, the high stress category for trees, shrubs and groundcover does not require any irrigation, but it rely only natural rainfall. In situations where irrigation is practiced, the irrigation interval must be sufficiently long to produce increasingly greater stress as soil water is depleted, and must cover a sufficiently long period so that the stress over the entire interval averages the desired value for ET_a/ET_p. The water stress over the time interval between irrigations ranges from 1.0, indicating no stress, over a period following irrigation (assuming that the irrigation depth was substantial) until the soil water depletion from the root zone exceeds *RAW*. Following that time, the ET_a/ET_p ratio will progressively decrease until the next irrigation. General values for K_L can be obtained from the above citations.

2.10 Estimating K_c from the Fraction of Ground Cover

When a K_c value is needed for vegetation that is not similar to that listed in Table 2.5, the K_d function of FAO-56 can be applied.

Table 2.11 Managed water stress (ET_a/ET_p) for general landscape plant types for no stress

Vegetation Category[1]	High stress	Average managed stress	Low stress
Trees	0.4	0.6	0.8
Shrubs species	0.3		
– desert species	0.4	0.4	0.6
– non desert		0.6	0.8
Groundcover	0.3	0.5	0.8
Annuals (flowers)	0.5	0.7	0.8
Mixture of trees, shrubs,	0.4		
and groundcover[2]		0.6	0.8
Cool season turfgrass[3]	0.7	0.8	0.9
Warm season turfgrass[4]	0.6	0.7	0.8

[1]The tree, shrub, and groundcover categories listed are for landscapes that are composed solely or predominantly of one of these vegetation types.
[2]Mixed plantings are composed of two or three vegetation types (i.e., where a single vegetation type does not predominate).
[3]Cool season grasses include Kentucky blue grass, fescues, perennial ryegrass.
[4]Warm season grasses include Bermuda grass, St. Augustine grass, buffalo grass, and blue grass.

The estimation of K_c for the initial growth stage of annuals, where the soil surface is mostly bare, can be determined for the mean K_c ($K_{c\ ini}$) using Fig. 2.10 where the crop coefficient in this stage is primarily determined by the frequency with which the soil is wetted. In the dual K_c approach, the K_{cb} for the initial period can be estimated as 0.1 to 0.15 for bare soil. The K_c during the midseason period ($K_{c\ mid}$), if for a low fraction of ground cover, will be affected to a large extent by the frequency of precipitation and/or irrigation and by the amount of leaf area and ground cover. Therefore, the basal $K_{cb} + K_e$ approach is recommended, with K_{cb} estimated using the following equation, according to the fraction of ground covered by vegetation during the particular period (see also Table 2.6):

$$K_{cb} = K_{c\ min} + K_d \left(K_{cb\ full} - K_{c\ min}\right) \tag{2.75}$$

where K_{cb} is the estimated basal K_{cb} (for example, during the midseason) when plant density and/or leaf area are at or below a full cover condition, $K_{c\ min}$ is the minimum K_{cb} representing bare soil, $K_{cb\ full}$ is the basal K_{cb} anticipated for the vegetation under full cover conditions and corrected for climate, and K_d is the density factor from Equation 2.77. Equation 2.75 can be used to estimate the single K_c, rather than K_{cb} by setting $K_{c\ min}$ equal to a '$K_{c\ ini}$ value' derived from Fig. 2.10 or from equations given in Allen et al. (1998, 2005c). For tree crops where ground cover may be present, Equation 2.65 can be expressed as:

$$K_{cb} = K_{cb\ cover} + K_d \left[max\left(K_{cb\ full} - K_{cb\ cover}, 0\right)\right] \tag{2.76}$$

where $K_{cb\ cover}$ is the K_{cb} of the ground cover in the absence of tree foliage.

Allen et al. (1998) introduced a general equation for K_d that is based on the fraction of ground covered (or shaded at noon) by vegetation and mean plant height. This relationship is shown in Table 2.12 and can be calculated using:

Table 2.12 Density factor (K_d) for different average vegetation heights, h, over a range of effective fraction of ground covered or shaded by vegetation, f_c, from Equation 2.77 where M_L = 1.5

f_c	h = 0.1 m	h = 0.4 m	h = 1 m	h = 4 m
0.0	0.00	0.00	0.00	0.00
0.1	0.12	0.15	0.15	0.15
0.2	0.23	0.30	0.30	0.30
0.3	0.33	0.42	0.45	0.45
0.4	0.43	0.52	0.60	0.60
0.5	0.53	0.61	0.71	0.75
0.6	0.63	0.69	0.77	0.90
0.7	0.72	0.78	0.84	0.93
0.8	0.82	0.85	0.89	0.96
0.9	0.91	0.93	0.95	0.98
1.0	1.00	1.00	1.00	1.00

$$K_d = min\left(1,\ M_L f_c,\ f_c^{\left(\frac{1}{1+h}\right)}\right) \qquad 2.77$$

Where f_c is the effective fraction of ground covered or shaded by vegetation (0 to 1.0) near solar noon, h is the mean height of the vegetation in m, and M_L is a coefficient that accounts for local (micro scale) advection between dry soil and canopy. Generally M_L = 1.5 for trees, shrubs and vine crops and M_L = 2.0 for vegetables and annual crops (see also Table 2.7). Equation 2.77 estimates larger values for K_d as vegetation height increases. This accounts for the impact of larger aerodynamic roughness and generally more leaf area with taller vegetation, given the same fraction of ground covered or shaded. The higher value for M_L for low growing crops accounts for more likelihood for dense annual vegetation near the ground level to capture horizontal micro scale transfer of heat. The 'min' function takes the smallest of the three values separated by the commas. The parameter $M_L f_c$ imposes an upper limit on relative magnitudes of transpiration per unit of leaf area as represented by f_c (Allen et al. 1998).

In Table 2.12, f_c is the fraction of ground covered or shaded by vegetation (0 to 1.0) near solar noon, h is the mean height of the vegetation in m, and M_L = 1.5. Table 2.12 estimates larger values for K_d as vegetation height increases. This accounts for the impact of larger aerodynamic roughness and generally more leaf area with taller vegetation, given the same fraction of ground covered or shaded.

2.11 Effects of Surface Mulching on K_c

Mulches are frequently used in vegetable production to reduce evaporation losses from the soil surface, to accelerate crop development in cool climates by increasing soil temperature, to reduce erosion, or to assist in weed control. Mulches may be composed of organic plant materials or they may be synthetic mulches comprised of plastic sheets. Plastic mulches are the most common type of mulch used in vegetable production.

2.11.1 Plastic Mulches

Plastic mulches are generally comprised of thin sheets of polyethylene or similar material placed over the ground surface and generally along plant rows. Holes are cut through the film at plant spacing's to allow emergence of vegetation. Polyethylene covers are usually either clear or black. Effects on ET_p by the two colors are generally similar (Haddadin and Ghawi 1983; Battikhi and Hill 1986a, b; Safadi 1991). Plastic mulches substantially reduce the evaporation of water from the soil surface, especially under trickle irrigation systems. Associated with the reduction in evaporation is a general increase in transpiration from vegetation caused by transfer of both sensible and radiative heat from the surface of the plastic cover to adjacent vegetation. Usually, the ET_p from mulched vegetables is about 5–30% lower than for vegetable production without plastic mulch. A summary of observed reductions in K_c, evaporation, and increases in transpiration over growing seasons is given in Table 2.13 for five vegetable crops. Even though the transpiration rates under mulch may increase by an average of 10–30% over the season as compared to using no mulch, the K_c value decreases by an average of 10–30% due to the 50–80% reduction in evaporation from wet soil. Generally, crop growth rates and vegetable yields are increased with the use of plastic mulches.

To consider the effects of plastic mulch on ET_p, the values for mean $K_{c\ mid}$ and $K_{c\ end}$ for vegetables listed in tables can be reduced by 10–30%, depending on the frequency of irrigation (use the higher value for frequent trickle irrigation). The value for $K_{c\ ini}$ is often as low as 0.10. When estimating basal K_{cb} for mulched production, less adjustment is needed to the K_{cb} curve, being on the order of perhaps 5–15% reduction in K_{cb}, since the "basal" evaporation of water from the soil surface is less with a plastic mulch, but the transpiration is relatively more. Local calibration of K_{cb} (and K_e) is encouraged. When applying a basal approach with plastic

Table 2.13 Approximate reductions in K_c, surface evaporation, and increases in transpiration for various vegetable crops under plastic mulch as compared with no mulch using trickle irrigation (From Allen et al. 1998)

Crop	*Reduction in* K_c[1] (%)	*Reduction in evaporation*[1] (%)	*Increase in Transpiration*[1] (%)
Squash	5–15	40–70	10–30
Cucumber	15–20	40–60	15–30
Cantaloupe	5–10	80	35
Watermelon	25–30	90	−10
Tomato	35	n/a	n/a
Average	10–30	50–80	10–30

[1] Relative to using no mulch.

mulch, f_w should represent the relative equivalent fraction of the ground surface that contributes to evaporation through the vent holes in the plastic cover. This fraction can be substantially (at least two to five times) larger than the area of the vent holes to account for vapor transfer from under the sheet.

2.11.2 Organic Mulches

Organic mulches are sometimes used with orchard production and row crops under reduced tillage operations. Organic mulches may be comprised of unincorporated plant residue or foreign material imported to the field. The depth of the organic mulch and the fraction of the soil surface covered can vary widely. These two parameters affect the amount of reduction in evaporation from the soil surface.

A general rule of thumb with a mulched surface is to reduce the amount of soil water evaporation by about 5% for each 10% of soil surface that is covered by the organic mulch. For example, if 50% of the soil surface were covered by organic crop residue mulch, then the soil evaporation would be reduced by about 25%. To apply this to the table K_c values, one would reduce $K_{c\ ini}$ values by about 25% and would reduce $K_{c\ mid}$ values by 25% of the difference between $K_{c\ mid}$ and $K_{cb\ mid}$.

When applying the basal approach with separate water balance of the surface soil layer, the magnitude of evaporation can be reduced by about 5% for each 10% of soil surface covered by the organic mulch. These recommendations are only approximate and attempt to account for the effects of partial reflection of solar radiation from residue, micro-advection of heat from residue into the soil, lateral movement of soil water from below residue to exposed soil, and the insulating effect of the organic cover. These parameters can vary widely, so that local research and measurement are encouraged.

Chapter 3
Effective Precipitation

3.1 Introduction

In order to estimate the irrigation water requirements, we first need to know how much of the soil water in the crop's rootzone will be provided by natural precipitation. Hence, precipitation needs to be measured (Fig. 3.1). Not all precipitation infiltrates into the soil; a part may evaporate; another part may become surface runoff. Of the precipitation that infiltrates, only a part will be stored in the root zone and the remainder will recharge the groundwater. Again, only a fraction of the total water stored (i.e. 'the readily available soil water') will be taken up by the roots to meet the crop's transpiration needs. Hence, when estimating the effective precipitation, we not only have to know the amount of actually depletable water, but also the fraction of the precipitation that becomes deep percolation and soil evaporation.

Effective precipitation is that part of the total precipitation that replaces, or potentially reduces, a corresponding net quantity of required irrigation water. We use the definition of effective precipitation that corresponds with the ICID terminology on the 'field application ratio' and the related water use efficiencies at crop production level (Bos 1980; Bos and Nugteren 1974; ICID 1978).

> *Effective precipitation is that part of the total precipitation on the cropped area, during a specific time period, which is available to meet the potential transpiration requirements in the cropped area.*

This definition limits itself to the 'cropped area'. Precipitation on fallow fields can range from very harmful to a future crop, to highly beneficial to it. Its value depends on a wide range of local conditions, which often discourages research on the effectiveness of precipitation. The phrase 'during a specific time period' may mean the entire period or any sub-period between sowing or planting and harvesting, or the period between harvests, which is decided upon from an agricultural or operational point of view.

The above definition limits the effective precipitation to that part 'which is available to meet evapotranspiration in the cropped area'. Precipitation which, upon infiltration, passes through the crop's rootzone may leach harmful salts from the soil. The rain may leach these salts during either a fallow period or the

M.G. Bos et al. *Water Requirements for Irrigation and the Environment*,

Fig. 3.1 The equipment used to measure precipitation influences the quantity measured where, the catch diameter, elevation, and surrounding ground cover all influence accuracy

Box 3.1 CRIWAR-quantified effective precipitation

To calculate the effective precipitation, CRIWAR uses two semi-empirical methods. In addition CRIWAR allows the user to set the effective precipitation as a fixed percentage of total precipitation. The three methods are:

- The method as developed by the U.S. Department of Agriculture (1970). This method can be used if monthly precipitation data is available. This method is described in Section 3.3.
- A method based on the Curve Number Method as developed by the U.S. Soil Conservation Service (1964 and 1972). This method requires daily precipitation data, as described in Section 3.4.
- The user sets P_e as a percentage of P, while P_e cannot exceed ET_p during the considered calculation period.

In using the USDA method (Section 3.3) or the curve number method (Section 3.4) it should be realized that these methods address two different processes. The USDA method estimates the part of the precipitation that is retained in the root zone for ET using infiltrated precipitation. There is no provision in the USDA method to estimate runoff. The Curve Number method, on the other hand, knows nothing and does nothing about estimating deep percolation. It only estimates runoff. Therefore, the best application is to combine these two methods. The CN method is first applied to estimate the part of precipitation that infiltrates, and then, that P_{inf} is used as the input to the USDA method.

crop season, or by non-consumed irrigation water. Water required for leaching serves a significant purpose, but is not included in the definition of effective precipitation. ICID proposed this definition so that data on effective precipitation, and the related field application ratio for different irrigated areas, could be compared without the errors by virtue of the local interpretation of the variable concept of the 'leaching water requirement'.

3.2 Major Factors Affecting Effective Precipitation

Various attempts have been made in order to establish a relationship between total precipitation and effective precipitation, either from individual storms or on a seasonal basis. Some methods use data on (cumulative) precipitation and evapotranspiration, soil data, and crop parameters to estimate the portion of the total precipitation that can be effective. The most sophisticated approaches are based on a dynamic simulation of a complete soil water balance on a day-to-day basis (Feddes et al. 1988; Kabat and Feddes 1989). Although these physically based dynamic models can provide very reliable information about the upper limit of the effective precipitation, they also need highly skilled users. Their application is therefore usually confined to sites where an extensive set of input data can be collected, but can be used to develop ratios of P_e to P to apply locally. We assume, however, that the irrigation manager does not have such detailed measured data.

Unfortunately, a universal formula relating the 'effective' to the total precipitation is not feasible because the ratio is affected by many independent factors, which will be discussed below. To appreciate the difference between the actual effective precipitation and the CRIWAR-quantified effective precipitation, one needs to have some basic knowledge of the major factors influencing this effectiveness. These factors are grouped in the flowchart in Fig. 3.2 (Kopec et al. 1984), which shows the path of measurable precipitation on an irrigated field. To follow this path, we have to make a number of decisions, which are shown as yes/no-exit blocks.

3.2.1 Amount and Frequency of Precipitation

In order to calculate irrigation water requirements, we have to interpret the precipitation data. Mean precipitation data are usually adequate for us to determine whether both the crop and the leaching requirements are being met in the long term (e.g. by precipitation during the non-irrigation season). To calculate the irrigation water requirement for, say, a 10 day period, we have to assess the amount of rainfall that can be reliably expected during that period. In this context, 'reliable' precipitation is usually taken to equal precipitation with a probability of between 80% and 90% (i.e. the amount of precipitation that will, on the average, be equalled or exceeded 80% or 90% of the time during this period).

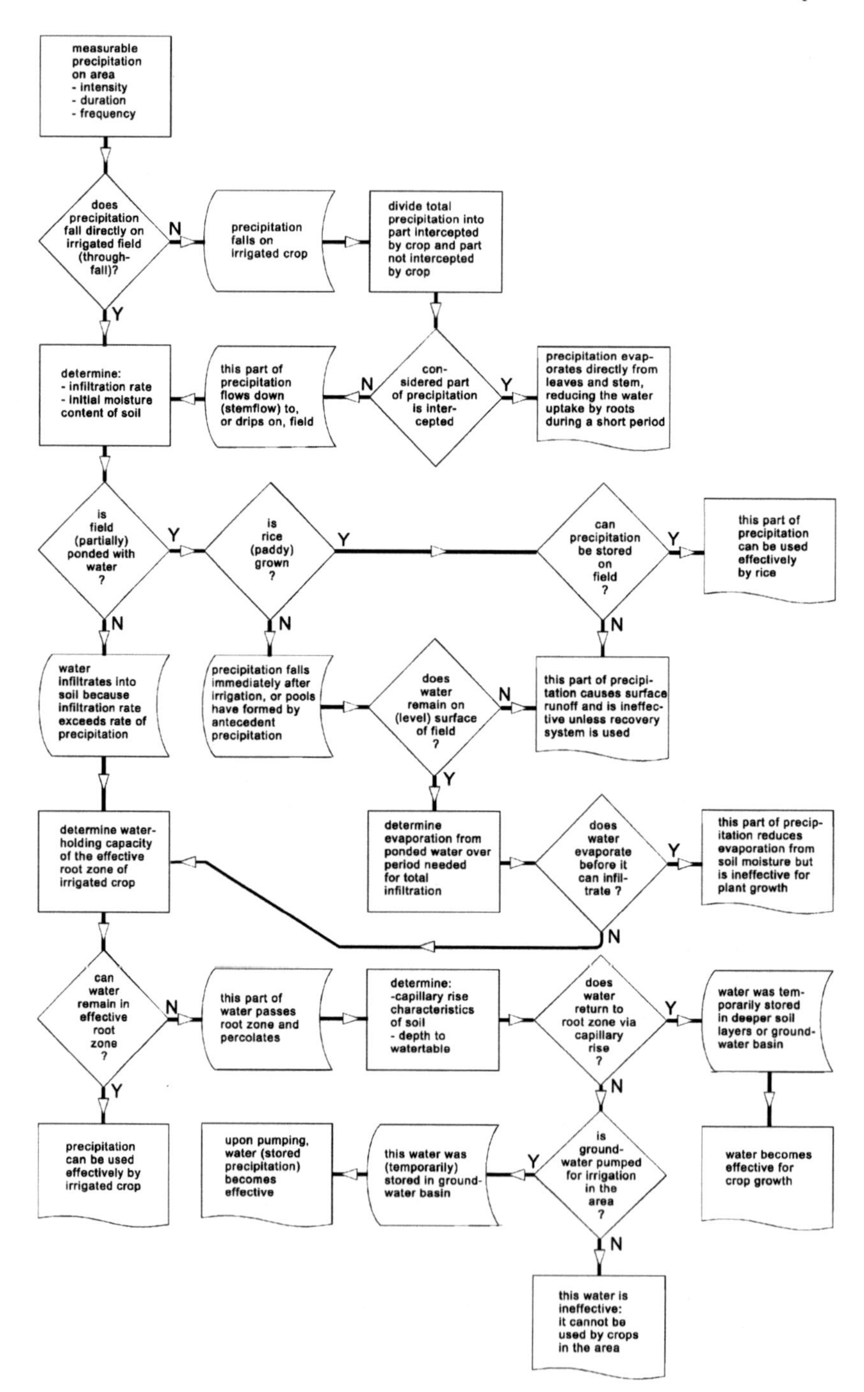

Fig. 3.2 Precipitation flow chart (Kopec et al. 1984)

3.2.2 *Time of Occurrence of Precipitation*

Snow or rainfall on frozen ground in the non-growing season usually runs off the land on which it falls and provides little or no soil moisture for later evapotranspiration. In contrast, snow on unfrozen ground may yield much of the soil moisture needed immediately after winter for a crop such as spring wheat.

If rain occurs immediately after an adequate application of irrigation water, all of it will theoretically be surplus to requirements and will thus be ineffective. At other times, between scheduled applications of irrigation water, the effectiveness of rainfall will be governed by the infiltration rate and the degree to which the soil moisture has been depleted below 'field capacity' (also see Section 3.4.2).

3.2.3 *Rainfall Intensity*

Due to runoff, an intensive downpour of, say, 100 mm in 1 h may yield much less effective precipitation than the same 100 mm spread over a longer period. Similarly, total precipitation of 100 mm spread over a series of very minor showers may evaporate from the sun-warmed surface without reaching the rootzone. In this case, precipitation is effective only to the extent that it:

- Replaces stored groundwater, which would otherwise have risen to the surface and evaporated
- Temporarily reduces the uptake of water through the roots
- The surplus, being evaporated, does not reduce the irrigation water requirement and is therefore not effective.

3.2.4 *Dry and Wet Spell Analysis*

In regions where rainfall is erratic, or where short dry periods can be expected in the wet season, we need to know the probability of occurrence of a dry period of 20 or 30 days. If such a dry period coincides with a sensitive stage in crop growth, yields will be reduced (Oldeman and Frere 1982). In order to avoid this yield reduction, supplemental irrigation water is often applied despite the fact that (erratic) rainfall may fall on the irrigated land.

3.2.5 *Irrigated Crops*

Shallow-rooting crops (e.g. onions and lettuce) require light but also frequent applications of water. In contrast, a heavy downpour will be much more effective for deep-rooted crops (e.g. sugarcane and alfalfa). The amount of evapotranspiration by a plant, and hence the requirement for irrigation water, also varies through the growing cycle. At some stages (e.g. in the early weeks or at the time of final ripening), only small amounts of soil moisture are removed per unit of time. With some crops, we may want to restrict evapotranspiration by reducing or stopping irrigation (e.g. to increase the sugar content of cane just before it is cut). Rainfall in sizeable quantities at these stages of plant growth will often be more than required and will therefore be less effective unless carried over as soil moisture for the next crop.

In general, it can be said that precipitation in excess of the storage capacity in the rootzone is ineffective. The USDA Soil Conservation Service (1970) developed a method to estimate the effective precipitation in which this rule is used. This method will be explained in Section 3.3.

3.2.6 Infiltration Rate

The infiltration rate is a characteristic of the soil and its state of preparation (e.g. whether it is tilled or not). Under otherwise identical conditions, a given shower of rain may be very effective if the soil has a sandy surface texture or has just been tilled. Conversely, it may be highly ineffective if the soil is an impermeable clay or has formed an impermeable caked surface; as most of the rainfall will either run off or be lost by evaporation from the surface. If lost by evaporation, it is ineffective to the extent that surface evaporation of standing water or wet soil surface exceeds losses of moisture that would have risen by capillary action (and evaporated) in the absence of rain (also see Section 3.4.2).

3.2.7 Water-Holding Capacity

The degree to which a soil is capable of holding or retaining moisture between field capacity and wilting point will limit the proportion of rainfall that will be held for subsequent use by the crop. A very coarse-textured sandy soil may allow rapid drainage through and beyond the rootzone to the underlying strata. A clay soil may be capable of retaining more moisture, provided that the rate of precipitation is sufficiently low in order to allow time for it to infiltrate into the rootzone.

3.2.8 Soil Water Movement

Much research has been done on the physics of soil water movement, and complex models have been developed to estimate the relationship between the soil ↔ plant ↔ atmosphere ↔ water. The most sophisticated approaches are based on a dynamic simulation of a complete soil water balance on a day-to-day basis (Feddes et al. 1988; Kabat and Feddes 1989).

3.2.9 Field Slope

The general slope of the ground surface is an important factor in determining the rate of application of irrigation water, particularly in furrow and border irrigation. Likewise, in intense storms or rapid snow melt, the slope of the ground will affect the degree of runoff and hence the proportion of water that will be retained in the soil. Precipitation on level basins with dry-foot crops may infiltrate entirely and in rice basins, part or all of the rainfall may be stored on the surface of the field.

3.2.10 Land Surface Condition

A dense crop cover will intercept precipitation and reduce the rate of runoff. This allows more time for the precipitation to infiltrate and thus increases the effective part of precipitation. The presence of surface mulches will also impede runoff and produce the same effect.

Tilling is an important factor. A hard, compacted surface reduces the rate of infiltration and increases runoff. Conversely, a well-tilled field will impede the surface flow of water and will increase both the infiltration rate and its duration. This can significantly increase the effectiveness of precipitation. Most soil conservation measures aiming at a reduction of runoff and an increase in the retention time of surface water have a similar effect. For example, the interception storage in agroforestry has a positive affect (also see Table 3.2).

3.2.11 Depth to Groundwater

The depth to groundwater has a major influence on the extent to which capillarity rise will bring groundwater into the rootzone or to the soil surface. If the plant roots extend into the capillary zone, much of the rain water that had previously percolated into the shallow groundwater remains in storage for subsequent evapotranspiration, thus increasing the effectiveness of previous precipitation. Capillary rise will be greater in fine-textured soils than in coarse-textured soils (also see Chapter 4).

3.2.12 Irrigation Water Supply Method

From the viewpoint of an irrigation system operator, there are various methods of supplying water to a group inlet of several small farms or to an individual inlet of one large farm:

- Continuous supply
- Rotational supply
- Supply on demand in advance or
- Supply on instantaneous demand

With the first two methods, only the statistically dependable part of the rainfall can be counted on to save irrigation water. Thus, although the other part of the rainfall may be highly effective in terms of the definition adopted, with these two water-supply methods, it does not permit a saving of irrigation water. With the third method, a saving of irrigation water in the event of effective rainfall is possible only if the unused irrigation water can remain in storage, or can be stored within the irrigation system for later use. The fourth water supply method allows the farmer to save irrigation water in the event of effective rainfall.

3.2.13 Frequency of Water Application

Frequency of irrigation water application and management of soil moisture is often ignored. Less frequent heavy irrigation will likely result in higher effective precipitation, than lighter frequent irrigation irrigations.

3.2.14 Scale Effect

Most of the aforementioned estimations on the magnitude of effective precipitation are for field scale. But at the irrigation system scale, rainfall and groundwater is sometimes recaptured within the irrigated area, making effective precipitation higher at that scale. This effect increases with the size of the area, since when considering areas over about 10,000 ha, a water balance study is recommended in order to improve the estimate of effective precipitation.

3.3 The USDA Method

In order to calculate the effective precipitation, a semi-empirical method developed by the U.S. Department of Agriculture (1970) can be used. This method is combined with an improved estimate of the effect of the net irrigation application depth on effective precipitation. On the basis of the information given in Section 3.2, the user is free either to reduce or to increase the estimate of the effective precipitation.

3.3.1 The Three Major Factors Used

The USDA method is based on a soil water balance performed for 22 meteorological stations in the U.S.A., by virtue of 50 years of data. It considers deep percolation to the groundwater basin and soil-profile depletion by evapotranspiration. Note that, surface runoff is only marginally accounted for in this method, and that three factors are considered to influence the effectiveness of precipitation. They will now be discussed below.

3.3.1.1 Mean Cumulative Monthly Precipitation

Rain storms of large magnitude and high intensity will supply water in excess of that which can be stored in the soil profile. Deep percolation to the groundwater and surface runoff will usually be high and in areas with light total precipitation during the growing season, these losses will not occur as frequently. As a consequence (by comparison) the effectiveness of precipitation in areas with light precipitation will be relatively high.

3.3.1.2 Mean Cumulative Monthly Evapotranspiration

When the evapotranspiration rate is high, the soil water will be rapidly depleted. As a consequence, a large amount of water can be stored in the soil profile again before it reaches field capacity. When the evapotranspiration rate is low, the storage capacity for precipitation will be provided at a slower rate. Thus, the higher the evapotranspiration rate, the higher the effectiveness of precipitation.

3.3.1.3 Irrigation Application Depth

For most irrigation areas, the depth of water application per irrigation turn is assumed to equal the readily available soil water that can be stored in the rootzone. The capacity of the soil profile to store water for crop use depends on the soil type and the effective rooting depth. Thus, a high storage capacity within the rootzone indicates a relatively high effectiveness of precipitation.

3.3.2 Calculation Method

Figure 3.3 shows the relationship between the three aforementioned factors. It shows that the average monthly effective precipitation can exceed neither the total average monthly rainfall nor the total evapotranspiration. For the same evapotranspiration rate, the effectiveness of precipitation, expressed as a percentage of the total precipitation, decreases with the higher total precipitation. The relationship in Fig. 3.3 is valid for a net irrigation water application depth of 75 mm per turn.

According to USDA, the effective precipitation is calculated on a monthly basis by an empirical expression which accurately describes the relationship in Fig. 3.3

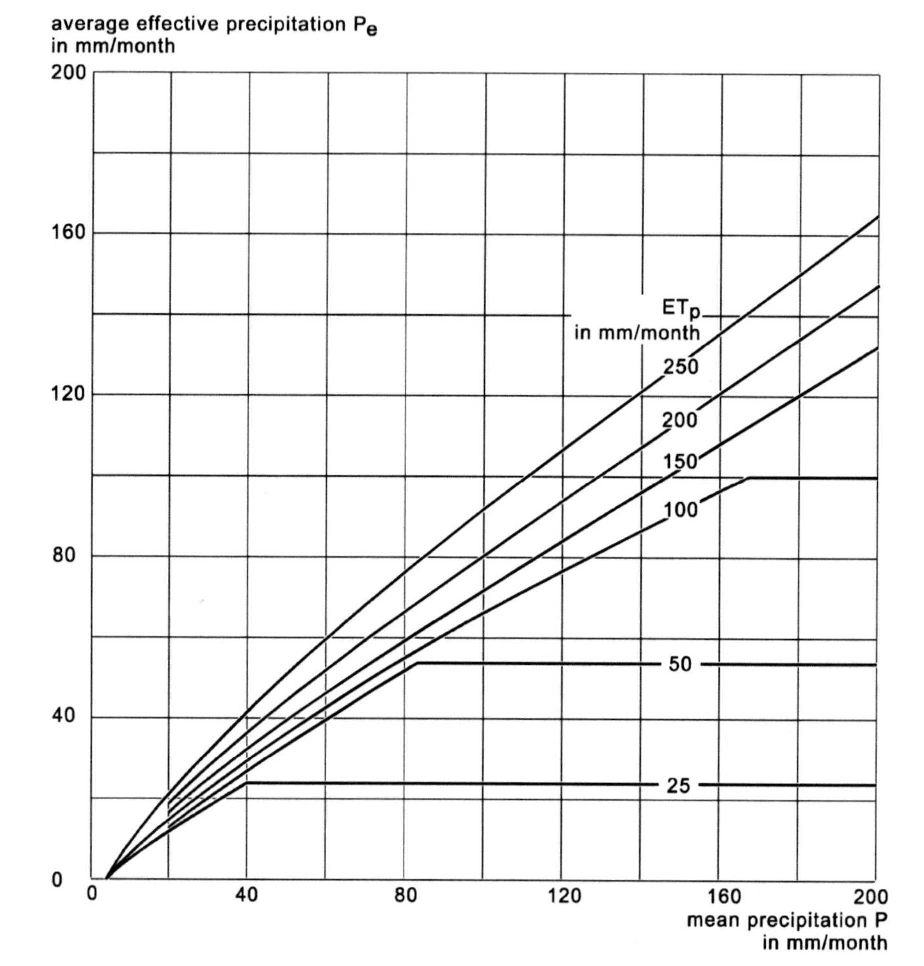

Fig. 3.3 Average monthly effective precipitation as related to the mean total monthly precipitation and the average monthly precipitation for a net irrigation application depth of 75 mm per turn

$$P_e = f \times (1.253P^{0.824} - 2.935) \times 10^{0.001ET_p} \qquad (3.1)$$

Where,
P_e = effective precipitation per month [mm/month]
P = total precipitation per month [mm/month]
ET_p = total crop evapotranspiration per month [mm/month]
f = a correction factor which depends on the depth of the irrigation water application per turn [-]
The factor f equals 1.0 if the irrigation water application depth is 75 mm per turn. For other application depths, the value of f equals:

$$f = 0.133 + 0.201 \ln D_a \qquad if \ D_a < 75\ mm\,/\,turn \tag{3.2}$$

and

$$f = 0.946 + 7.3 \times 10^{-4} \times D_a \qquad if \ D_a \geq 75\ mm\,/\,turn \tag{3.3}$$

If the use of these equations results in an effective precipitation that exceeds either ET_p or P, CRIWAR reduces the P_e value to the lowest of these two. When the mean total rainfall per month is less than 12.5 mm, CRIWAR assumes all precipitation to be 100% effective.

If a calculation per day, week or every 10-days is requested, CRIWAR converts the user-given precipitation data into total monthly data. Also the calculated ET_p is converted to total monthly crop evapotranspiration. With these converted data, the effective rainfall is estimated from Equations 3.1–3.3. After that, the calculated effective precipitation in mm/month is converted back into mm/day, mm/week or mm/10-days.

3.4 The Curve Number Method

3.4.1 Background of the Method

The Curve Number method was originally developed by the Soil Conservation Service (SCS 1964, 1972) in order to estimate the depth of direct surface runoff from the precipitation depth. The method was developed to be used with daily precipitation data measured with (non-recording) rain gauges. The relationship therefore excludes time as an explicit variable (i.e. rainfall intensity is not included in the estimate of runoff depth). CRIWAR uses the CN Method in order to estimate the part of precipitation that can be used to meet evapotranspiration from the (irrigated) field.

Figure 3.4 shows the rate of precipitation as a function of time (days). Following the start of precipitation, the first water will be intercepted by the crop, stored in small depressions, and infiltrate in the soil (initial abstraction, I_a). After runoff has started, all additional precipitation becomes either runoff (Q) or actual retention (F). The sum of I_a and F becomes available for evapotranspiration from the (cropped) field.

The 'CN method' is named after the plotted curve of accumulated precipitation and runoff of long duration over a small drainage basin (Fig. 3.5). The curve shows that runoff (Q) only starts after some precipitation has accumulated and that the curve asymptotically approaches a straight line with a 45-degree slope. This initial accumulation (I_a) represents interception by the crop, depression storage and infiltration. After runoff has started, some of the additional precipitation will infiltrate; this part is called actual retention (F). With increasing precipitation, the actual retention also increases up to some maximum value: the potential maximum retention (S).

To describe the curves with the shape of Fig. 3.5, the SCS assumed that the ratio of actual retention to potential maximum retention was equal to the ratio of actual

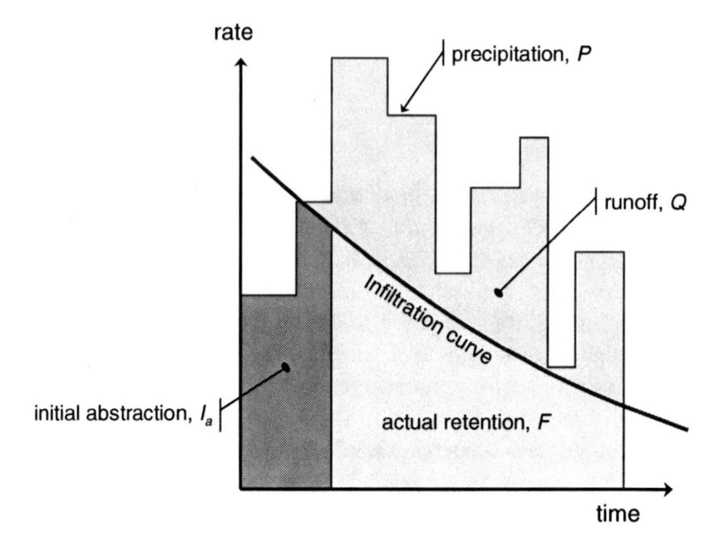

Fig. 3.4 The division of precipitation between initial abstraction, actual retention, and runoff

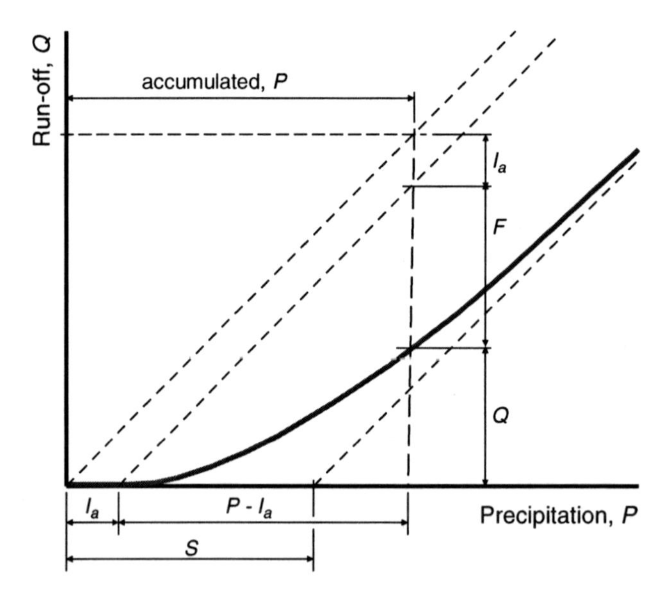

Fig. 3.5 Accumulated runoff versus accumulated precipitation according to the Curve Number Method

runoff to potential maximum runoff, the latter being precipitation minus initial abstraction. In mathematical form, this empirical relation is:

$$\frac{F}{S} = \frac{Q}{P - I_a} \tag{3.4}$$

Where,
F = actual retention (mm)
S = maximum potential difference between precipitation and runoff beginning at the time precipitation starts (also named maximum retention) (mm)
Q = accumulated runoff depth (mm)
P = accumulated precipitation depth (mm)
I_a = initial abstraction (mm)
Figure 3.5 shows the above relationship. After runoff has started, all additional precipitation becomes either runoff or actual retention (neglecting actual ET). Thus:

$$F + I_a = P - Q \tag{3.5}$$

For irrigation purposes we assume that $F + I_a$ equals the effective precipitation, P_e. Therefore, combining Equations 3.4 and 3.5 yields,

$$Q = \frac{(P - I_a)^2}{P - I_a + S} \tag{3.6}$$

In order to eliminate the need to estimate two retention variables (i.e. I_a and S) in Equation 3.6, a regression model was made on the basis of recorded rainfall runoff data from small drainage basins. As the data showed a large amount of scatter (SCS 1972), the following average relationship was found:

$$I_a = 0.2\,S \tag{3.7}$$

Combining Equations 3.6 and 3.7 yields

$$Q = \frac{(P - 0.2S)^2}{P + 0.8\,S} \qquad for\ P \geq 0.2\,S \tag{3.8}$$

Equation 3.8 is the rainfall-runoff relationship used in the Curve Number Method. It allows the runoff depth to be estimated from rainfall depth, if the value of the potential maximum retention, S is known. Equations 3.7 and 3.8 imply that $Q = 0$ if $P \leq 0.2S$. Thus, all precipitation would be available to meet actual evapotranspiration and is quantified as effective precipitation.

As mentioned in Section 3.2 the value of S is controlled by the rate of infiltration at the soil surface, or by the rate of transmission in the soil profile, or by the water-storage capacity of the profile, whichever is the limiting factor. The potential maximum retention S has been converted to the Curve Number, CN in order to make the operations of interpolation, averaging, and weighting more linear. This relationship is defined as:

$$\text{CN} = \frac{25400}{254 + S} \tag{3.9}$$

As the value of S can theoretically vary between zero and infinity, Equation 3.9 shows that the Curve Number, CN can range from 100 to 0. Figure 3.6 shows the

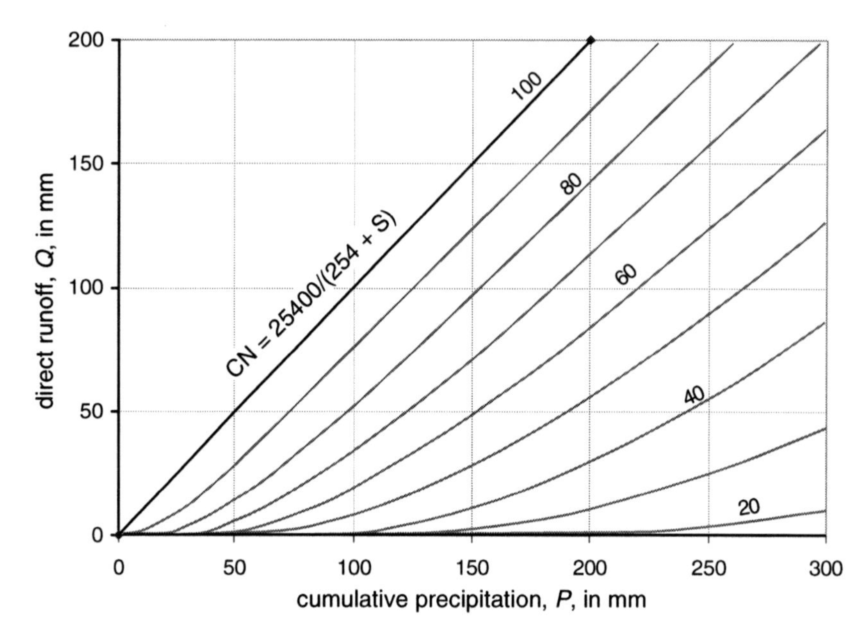

Fig. 3.6 Graphical solution of Equation 3.8 showing runoff depth Q as a function of precipitation P and the curve number CN (After SCS 1972)

graphical solution of Equation 3.6, indicating values of runoff depth Q as a function of precipitation depth, P for selected values of Curve Numbers.

CN = 0 No direct runoff	CN = 100 All precipitation runs off
S = very large All precipitation infiltrates	S = 0 No infiltration

For example, when considering very heavy sloping clays, hardly any precipitation will infiltrate, and S will approach zero. Thus, Equation 3.9 shows that CN will be 100, i.e. all precipitation will become runoff. For highly permeable, flat-lying soils, all precipitation can infiltrate and S will go to infinity. Thus, CN will be zero and there will be no runoff. In irrigated areas, the reality will be somewhere in between.

3.4.2 Factors Determining the Curve Number Value

The Curve Number is a dimensionless parameter indicating the runoff response characteristics of an area. In the Curve Number Method, this parameter is related to land use, land treatment, hydrological condition, hydrological soil group, and antecedent soil moisture of the area.

3.4.2.1 Land Use or Cover

Land use represents the surface conditions of the area and is related to the degree of cover. In the CNM the following categories are distinguished (Table 3.1):

Fallow is the agricultural land use with the highest potential for runoff because the land is kept bare.

Row crops are field crops panted in rows far enough apart that most of the soil surface is directly exposed to precipitation.

Table 3.1 Curve number values for the combined 'Hydrological Soil Cover Complex' for average antecedent soil moisture (class II), flat or slightly sloping areas, and $I_a = 0.2S$ (SCS 1972)

			Hydrological soil group			
Land use or cover	Treatment or practice	Hydrological conditions	A	B	C	D
Fallow	Straight row	Poor	77	86	91	94
Row crops	Straight row	Poor	72	81	88	91
	Straight row	Good	67	78	85	89
	Contoured	Poor	70	79	81	88
	Contoured	Good	65	75	82	86
	Terraced	Poor	66	74	80	82
	Terraced	good	62	71	78	81
Small grain	Straight row	Poor	65	76	84	88
	Straight row	Good	63	75	83	87
	Contoured	Poor	63	74	82	85
	Contoured	Good	61	73	81	84
	Terraced	Poor	61	72	79	82
	Terraced	Good	59	70	78	81
Close-seeded legumes	Straight row	Poor	66	77	85	89
Or rotational meadow	Straight row	Good	58	72	81	85
	Contoured	Poor	64	75	83	85
	Contoured	Good	55	69	78	83
	Terraced	Poor	63	73	80	83
	Terraced	Good	51	67	67	80
Pasture range		Poor	68	79	86	89
		Fair	49	69	79	84
		Good	39	61	74	80
	Contoured	Poor	47	67	81	88
	Contoured	Fair	25	59	75	83
	Contoured	Good	6	35	70	79
Meadows (permanent)		Good	30	58	71	78
Woodlands		Poor	45	66	77	83
(farm woodlots)		Fair	36	60	73	79
		Good	25	55	70	77
Farmsteads			59	74	82	86
Roads, dirt			72	82	87	89
Roads (hard surface)			74	84	90	92

Small grain is planted in rows close enough that the soil is not directly exposed to precipitation.

Close-seeded legumes or rotational meadow are either planted in close rows or broadcasted. This kind of crop usually protects the soil throughout the year.

Pasture range is native grassland used for grazing, whereas meadow is grassland protected from grazing and generally mown from hay.

Woodlands are usually small isolated groves of trees being raised for farm use (orchards).

3.4.2.2 Treatment or Practice in Relation to Hydrological Condition

Land treatment applies mainly to agricultural land uses. It includes mechanical practices such as contouring or terracing, and management practices such as rotation of crops, grazing control, or burning. The ease with which water can infiltrate the soil is rated as: good, fair or poor, as follows (also see Table 3.1):

Crop rotations; poor rotations are generally one-crop land uses (monoculture) or combinations of row crops, small grain and fallow. Good rotations generally contain close-seeded legumes or grass.

For grazing control and burning (pasture range and woodlands), the hydrological condition is classified as poor, fair or good. Pasture range is classified as poor when heavily grazed and less than half the area is covered; as fair when not heavily grazed and between one-half to three-quarters of the area is covered; and as good when lightly grazed and more than three-quarters of the area is covered.

Woodlands are classified as poor when heavily grazed or regularly burned; as fair when grazed but not burned; and as good when protected from grazing.

3.4.2.3 Hydrological Soil Group

Soil properties greatly influence the amount of infiltration and runoff. In the SCS method, these properties are represented by a hydrological parameter: the minimum rate of infiltration obtained for a bare soil after prolonged wetting. The influence of both the soil's surface condition (infiltration rate) and its horizon (transmission rate) are thereby included. This parameter, which indicates a soil's runoff potential, is the qualitative basis of the classification of all soils into four groups. The Hydrological Soil Groups (as defined by SCS scientists) are:

Group A:	Soils having high infiltration rates even when thoroughly wetted and a high rate of water transmission. Examples are deep, well to excessively drained sands or gravels.	Final infiltration rate 8–12 mm/h
Group B:	Soils having moderate infiltration rates when thoroughly wetted and a moderate rate of water transmission. Examples are moderately deep to deep, moderately well to well drained soils with moderately fine to coarse texture.	Final infiltration rate 4–8 mm/h

Group C:	Soils having low infiltration rates when thoroughly wetted and a low rate of water transmission. Examples are soils with a layer that impedes the downward movement of water or soils of moderately fine to fine texture.	Final infiltration rate 1–4 mm/h
Group D:	Soils having a very low infiltration rate when thoroughly wetted and a very low rate of water transmission. Examples are: clay soils with a high swelling potential, soils with a permanent high watertable, soils with a clay pan or clay layer at or near the surface, or shallow soils over nearly impervious material.	Final infiltration rate less than 1 mm/h

3.4.2.4 Antecedent Moisture Conditions

The soil moisture condition in the area before runoff occurs is another important factor influencing the final CN value. In the Curve Number Method, the soil moisture condition is classified in three Antecedent Soil Moisture Conditions (AMC) classes:

AMC I: The soils in the area are practically dry (i.e. the soil moisture content is at wilting point).
AMC II: Average conditions.
AMC III: The soils in the area are practically saturated from antecedent precipitation or irrigation water application (i.e. the soil moisture content is at field capacity).

These classes are based on the 5 day antecedent precipitation and irrigation (i.e. the accumulated total water received preceding the runoff under consideration). In the original SCS method, a distinction was made between the dormant and the growing season in order to allow for differences in actual evapotranspiration (Table 3.2).

3.4.3 Estimating the Curve Number Value

In order to determine the appropriate CN value, the impact of various factors on the division of precipitation between runoff and infiltration has to be estimated. For this, the SCS (1972) made Table 3.1 relating the CN value to land use cover, to treatment or practice, to hydrological condition, and to the hydrological soil group. The CN values of Table 3.1 are valid for flat or slightly sloping areas, for an average

Table 3.2 Seasonal precipitation (plus irrigation) limits for AMC classes (After SCS 1972)

Antecedent moisture condition class	5-day antecedent rainfall plus irrigation application (mm/5 days)	
	Dormant season Growing season	Growing season
AMC I	<13	<36
AMC II	13–28	36–53
AMC III	>28	>53

relationship of $I_a = 0.2S$ (see Equation 3.7), and for average antecedent soil moisture (AMC II of Table 3.2). If the 5 day antecedent precipitation plus irrigation application is classified as either Class I or Class III, the CN value of Table 3.1 needs to be corrected with an appropriate factor, as given in Table 3.3.

Example:
The Tadla irrigation district is located in Morocco, at latitude 32° 28′ north and at an altitude of 434 m above mean sea level (see CRIWAR general data file). Precipitation is measured daily at several meteorological stations (Fig. 3.7).

In order to select a suitable CN value from Table 3.1, information is needed on; land use, treatment practice, and the hydrological soil group. For the example area this is:

- The land cover in the irrigated area consists of row crops (sugar beets).
- The sugar beets rotate with other irrigated row crops and temporary fallow. The hydrological condition due to this treatment is classified as poor.
- The soils in the area are moderately well to well drained, and have a moderately fine to coarse texture. The hydrological soil group is 'B'.

Analysing Table 3.1 yields CN = 81. As mentioned before, this CN value (from Table 3.1) is valid for flat or slightly sloping areas, and for average antecedent soil moisture (AMC II of Table 3.2). The distribution of precipitation between Q and $(F + I_a)$ can now be estimated for each daily value of P as follows:

Table 3.3 Conversion table for Curve Number values from antecedent moisture class II to class I and to class III (SCS 1972)

Value from Table 3.2	Corrected for AMC I and III		Value from Table 3.2	Corrected for AMC I and III	
AMC II	AMC I	AMC III	AMC II	AMC I	AMC III
100	100	100	56	36	75
96	89	99	54	34	73
92	81	97	50	31	70
90	78	96	48	29	68
86	72	94	44	25	64
84	68	93	42	24	62
80	63	91	38	21	58
78	60	90	36	19	56
74	55	88	32	16	52
72	53	86	30	15	50
68	48	84	20	9	37
66	46	82	15	6	30
62	42	79	5	2	13
60	40	78	0	0	0

Fig. 3.7 Meteorological station, Tadla, Morocco

Day 115 in Table 3.4

Prior to the first precipitation on day 114, a depth of 40 mm irrigation water was applied. According to Table 3.2 this yields an antecedent soil moisture class II (36–53 mm during the growing season). The CN-value thus needs no correction for the antecedent soil moisture. Substituting CN = 81 into a reshaped version of Equation 3.9 gives:

$$S = \frac{25400}{\mathrm{CN}} - 254 = \frac{25400}{81} - 254 = 60\mathrm{mm} \tag{3.10}$$

Substituting $S = 60$ mm into Equation 3.8 gives:

$$\mathrm{Q} = \frac{(\mathrm{P} - 12)^2}{\mathrm{P} + 48} \quad \text{for P} > 12\,\text{mm/day, (Q = 0 otherwise)} \tag{3.11}$$

For $P = 11$ mm/day, Equation 3.11 gives $Q = 0$ mm/day. Substitution of this value into Equation 3.5 gives (see Table 3.4):

$$F + I_a = 11 - 0 = 11\,\mathrm{mm/day} \tag{3.12}$$

Day 116 in Table 3.4

The 5 day antecedent precipitation and irrigation for Day 116 is 40 + 11 = 51 mm. Thus, the AMC class remains II. Hence, for day 116, S remains 60 mm. Substitution of P = 24 mm/day into Equation 3.11 gives Q = 2 mm/day. Thus $P_e = F + I_a = 24 - 2 = 22$ mm/day.

Table 3.4 Actual precipitation together with the preceding irrigation water application (mm/day)

Julian day	Actual precipitation P	Applied irrigation water	Antecedent soil moisture class	CN	Effective precipitation* $P_e = F + I_a$
113	0		II	81	0
114	0	40	II	81	0
115	11		II	81	11
116	24		II	81	22
117	32		III	92	16
118	8		III	92	8
119	0		III	92	0
120	0		III	92	0
121	0		II	81	0
122	0		II	81	0
123	27		I	64	27
124	0	20	I	64	0

**CRIWAR will reduce these values in such a way that the sun of the effertive precipitation over the next 30 days (month) is less than the moving potential evapotranspiration over this month*

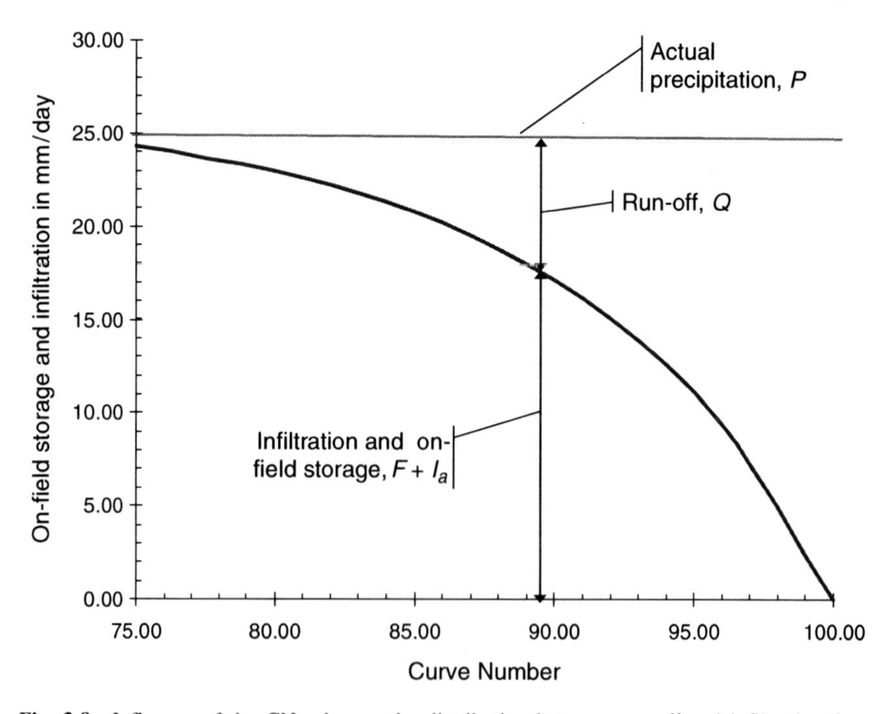

Fig. 3.8 Influence of the CN-value on the distribution between run-off and infiltration for a precipitation of 25 mm/day

Day 117 in Table 3.4

The 5-day antecedent precipitation and irrigation for Day 117 is 40 + 11 + 24 = 75 mm. Thus, the AMC class changes to III (see Table 3.2). Using Table 3.3 we find a CN-value of 92 for class III. Substitution of CN = 92 into Equation 3.9 gives S = 22 mm. Substitution of P = 32 mm/day into Equation 3.11 gives Q = 16 mm/day. Thus $P_e = F + I_a = 32 - 16 = 16$ mm/day.

Table 3.4 has been completed for other days. The above example shows that the selected CN-value and the antecedent soil moisture have a huge impact on the distribution of precipitation between Q and $(F + I_a)$. In order to illustrate this impact, Fig. 3.8 shows this distribution for a Precipitation of 25 mm/day. As shown, $P_e = (F + I_a)$ ranges from 24 mm/day to 0 if CN moves from 75 to 100. However, in order to improve the estimate of CN, the run-off should be measured for the drainage basin in which the meteorological data are measured. The use of a long-throated flume is recommended for the measurement of run-off (Clemmens et al. 2001).

Chapter 4
Capillary Rise

4.1 Introduction

Figure 4.1 shows the water balance of an irrigated field. As illustrated, the crop receives water through precipitation, P (see Chapter 3 for the related effective precipitation, P_e), through irrigation, and through capillary rise. If the depth to the groundwater table is shallow (less than about 3 m) and the soil is fine-textured, capillary rise can contribute a significant volume of water to the rootzone of the crop. However, in order for the groundwater table to remain stable, groundwater must flow laterally into the irrigated area; otherwise the capillary rise will decrease with the falling groundwater table.

Although groundwater flow is not simulated in CRIWAR, the capillary component is corrected for in the irrigation water requirements. This chapter illustrates the case of when capillary rise is a potential source of water. In which case, the calculated crop water requirements should be corrected for the contribution from groundwater (see Section 4.6).

4.2 The Driving Force of Capillary Water

Soil can be regarded as a mixture of solids and pores, with the pores forming capillary tubes. If the bottom end of a capillary tube is inserted in water, the water will rise into the tube under the influence of capillary forces (Fig. 4.2). The total upward force, $F\uparrow$, lifting the water column, is obtained by multiplying the vertical component of surface tension by the circumference of the capillary tube,

$$F\uparrow = \sigma \cos\alpha \times 2\pi r \qquad 4.1$$

Where,
$F\uparrow$ = upward force (N)
σ = surface tension of water against air (σ = 0.073 kg s^{-2} at 20°C)
α = contact angle of water with the tube (rad); (cos α = 1.0)

M.G. Bos et al. *Water Requirements for Irrigation and the Environment*,

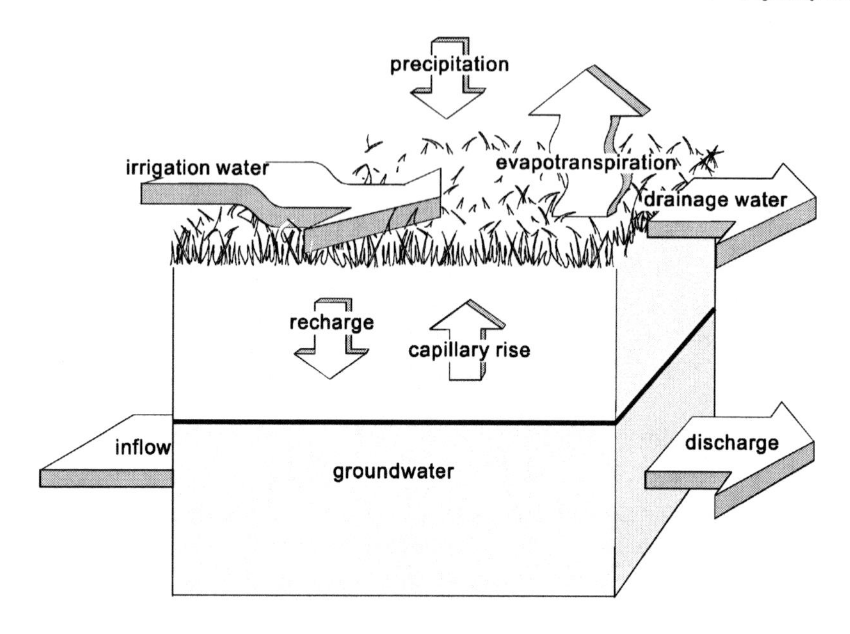

Fig. 4.1 The water balance of an irrigated field (Bos 1984)

r = equivalent radius of the tube (m)

Due to the force of gravity, the water column of height C and mass $\pi r^2 C\rho$ exerts a downward force, $F\downarrow$, that opposes the capillary rise

$$F\downarrow = \pi r^2 C\rho \times g \qquad 4.2$$

where
$F\downarrow$ = downward force (N)
ρ = density of water (ρ = 1,000 kg/m^3)
g = acceleration due to gravity (g = 9.81 m/s^2)
C = height of capillary rise (m)

At equilibrium, the upward force, $F\uparrow$, must equal the downward force, $F\downarrow$. Hence, $\sigma \cos\alpha \times 2\pi r = \pi r^2 C\rho \times g$

or

$$C = \frac{2\sigma \cos\alpha}{\rho g r} \qquad 4.3$$

Substituting the values for σ, cos α, ρ, and g as given above into Equation 4.3, gives an expression for the height of the capillary rise:

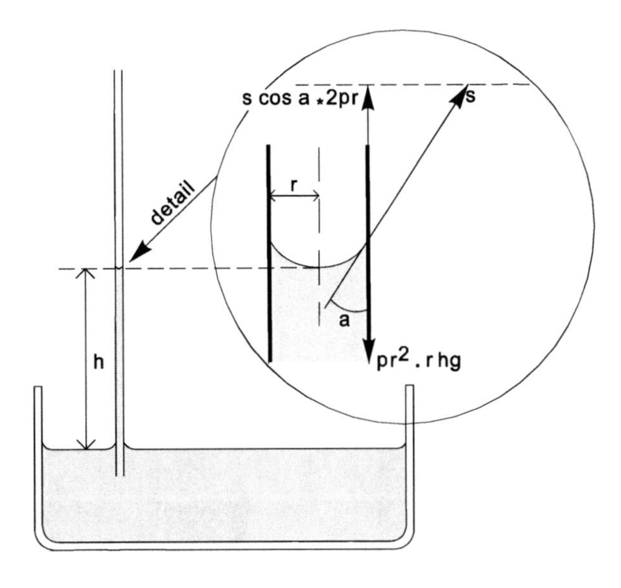

Fig. 4.2 Capillary rise of water (Kabat and Beekma 1994)

$$C = \frac{0.15}{r} \quad 4.4$$

Thus, the smaller the radius of the tube, the higher the capillary rise. However, real soils do not consist of capillaries of one uniform diameter, and furthermore, water movement in real soil is influenced by thermal, electrical, and solute-concentration gradients. However, for our purposes let us assume that an elementary water particle has three types of interchangeable energy per unit of volume:

$\rho v^2/2$ = kinetic energy per unit of volume (Pa)
ρgz = potential energy per unit of volume (Pa)
p = pressure energy per unit of volume (Pa)

The flow velocity of water in the soil pore is very low, so $\rho v^2/2$ is negligible. If the other two energies of water are divided by ρg, the hydraulic energy head, h, can be written as:

$$h = \frac{p}{\rho g} + z \quad 4.5$$

The above pressure head, $p/\rho g$, is negative in unsaturated soil because energy is needed to withdraw water against the soil-matric forces. At the groundwater level, atmospheric pressure exists and therefore $p/\rho g = 0$.

The elevation head, z, is determined at each point by the elevation of that point relative to a certain reference level (z being positive above the reference level and negative below it – see Fig. 4.3).

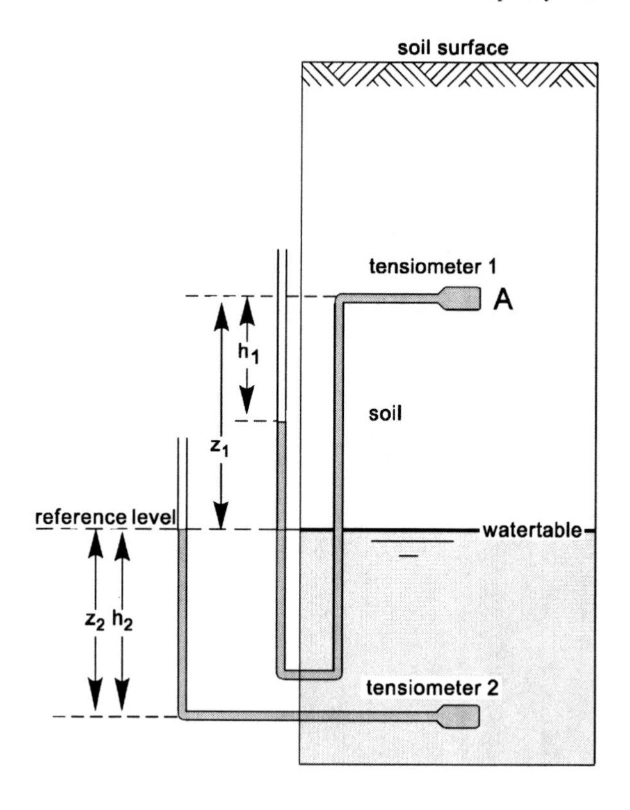

Fig. 4.3 The hydraulic head, h, at point A, located at a height, z, above a reference level

4.3 Steady-State Capillary Rise

The most simple case of capillary flow is that of steady-state vertical flow in an isotropic media (i.e. a soil whose hydraulic conductivity is the same in every direction). The flow equation is obtained by rewriting Darcy's Equation:

$$q = -K\left(\frac{dh}{dz} + 1\right) \tag{4.6}$$

Where,
q = vertical flow rate per unit area (m/day)
K = hydraulic conductivity as a function of h (m/day)
h = (hydraulic) head (m)
z = elevation head, being positive in the upward direction (m)

Rearranging Equation 4.6 yields:

$$\frac{dz}{dh} = -\frac{1}{1 + \frac{q}{K}} \tag{4.7}$$

In order to calculate the (hydraulic) head distribution (i.e. the relationship between z and h for a certain K-relationship and a specified flow rate q), Equation 4.7 should be integrated. This yields:

$$\int_0^C dz = -\int_0^{h_{pr}} \frac{dh}{1+\frac{q}{K}} \qquad 4.8$$

Where,
h_{pr} = the pressure head, $p/\rho g$, at the upper boundary condition (m)
C = the height of capillary rise for flow rate q (m)
If the hydraulic head (thus also h_{pr}) and the hydraulic conductivity are measured in the soil profile as a function of elevation (head), Equation 4.8 can be solved by integration between the head at the groundwater level ($h = 0$ m) and the measured value of h_{pr} for constant values of q. For complex combinations of these parameters, Equation 4.8 (one for each soil layer) can be solved by numerical models (Wesseling 1991; Raes and deProost 2003).

As mentioned earlier, the capillary flux (flow rate per unit area) and the capillary rise above the groundwater table both depend on the soil type and on the pressure differential between the groundwater table and the upper boundary condition. If, for example, the soil surface is the upper boundary, and if we assume a stable groundwater table and a gradual increase in the soil pressure head from zero at the groundwater level to a value of $h_{pr} = -160$ m ($pF = 4.2$) at the upper boundary (a value corresponding with a soil water content at wilting point), we can calculate the relationship between C and q. Figure 4.4 shows this relationship for four undisturbed Dutch soils (Wösten 1987). Thus, Fig. 4.4 illustrates several interesting points:

- Upward flow rates of more than 2 mm/day are common. Nevertheless, the height over which this flow can rise depends on the soil type (and structure), but even for coarse sand, this height is still about 0.4 m.
- The 'maximum' height of capillary rise in heavy clay is much greater than that in coarse sand. However, because of the low hydraulic conductivity of clay, the flow rate is low.
- If the lower side of the effective rooting depth is near the groundwater table (say <0.5 m), the groundwater contribution to crop water consumption is considerable.
- In fine-textured soils, the capillary flow rate varies more with the height than in coarse-textured soils.

If the groundwater table remains at a constant level as a result of the lateral inflow of groundwater, the capillary flow rate will remain constant. However, if the groundwater is not fed by lateral inflow, the capillary flow will cause the groundwater table to fall. As shown in Fig. 4.4, with a greater depth to the groundwater table, the capillary flow decreases sharply. The end result is that the groundwater table falls to a depth where the capillary flow rate is zero.

As mentioned earlier, Fig. 4.4 is based on the assumption that the soil moisture pressure at the lower side of the effective rootzone is $h_{pr} = -160$ m (wilting point). Following an irrigation water application, soil moisture will increase and the

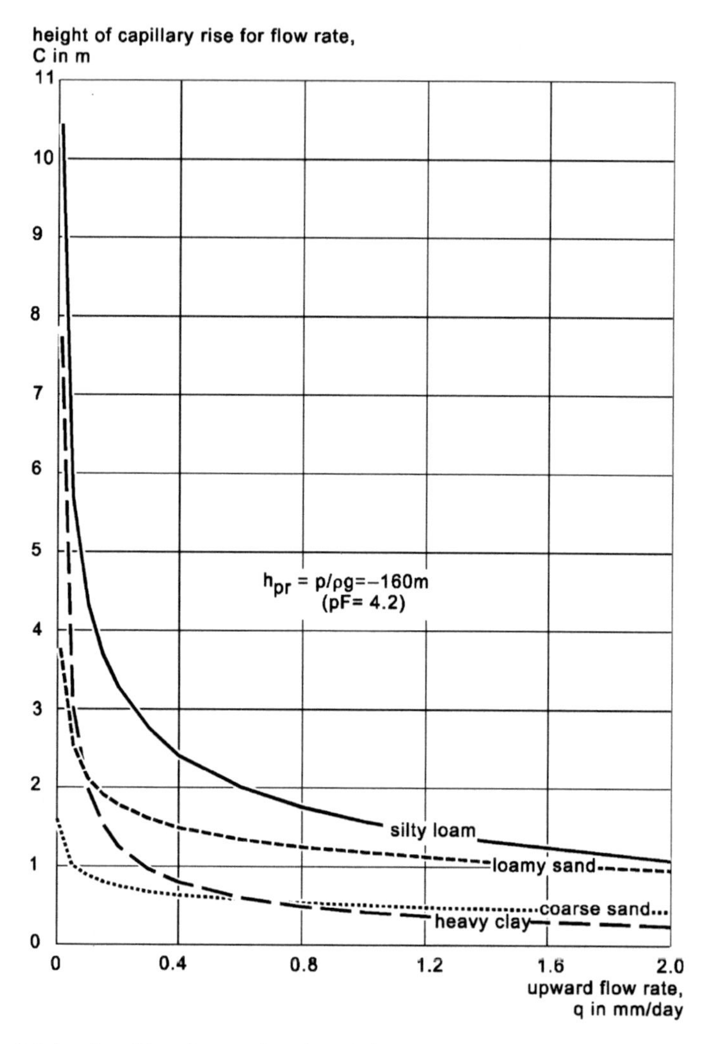

Fig. 4.4 Height of capillary rise as a function of the upward flow rate (flux) for four undisturbed Dutch soils

capillary flow rate (flux) will decrease with the decrease in h_{pr}. As a result of variable soil moisture, the depth to the groundwater table, and the effective rooting depth, the capillary contribution of water to crop consumption varies with time.

4.4 Fluctuation of Groundwater Depth and Soil Moisture

The water balance of an irrigated area shows three sources of water: precipitation, groundwater inflow and river (surface) water diversion (Fig. 4.1). Part of all this water evapo-tranpires, and the remainder will flow to downstream areas either via

Fig. 4.5 Fields with different balance between (annual) seepage and capillary rise. On the right hand side seepage exceeds capillary rise by about 15%

surface streams (drains) or as groundwater. If the summed inflow exceeds the outflow, part of the water will be stored in the unsaturated zone and in the aquifer within the irrigated area. If, on the other hand, the summed outflow exceeds the inflow water will be taken from storage. As shown in Section 4.3, the resulting drop of the groundwater table will reduce the availability of capillary water to crop growth. With a continued lowering of the groundwater table this capillary water resource will not be available for the crop. From a water management point of view, the 'non-availability' of capillary water requires a more accurate supply of irrigation water (Fig. 4.5).

For example, if the ET_p is 6 mm/day and the upward capillary rise is 1.5 mm/day, irrigation should deliver the missing 4.5 mm/day. If this water delivery is too low or too late (dry period, repair of infrastructure, etc.) the crop would feel water stress. However, it could survive several days on the capillary rise. Hence, it is recommended to manage irrigation water (V_c) in such a way that the groundwater remains a reliable source of water. For this to occur, the groundwater table should remain stable from year to year (Section 4.6). The example of Fig. 4.6 shows a gradually rising groundwater table that fluctuates heavily throughout the year. The gradual rise eventually may cause water logging (and salinity) while the fluctuation complicates the forecast of the availability of capillary water.

As discussed in Section 1.3, the rate of change of soil moisture in the unsaturated zone and the rate of change of the groundwater table are influenced by the depleted fraction of the gross command area:

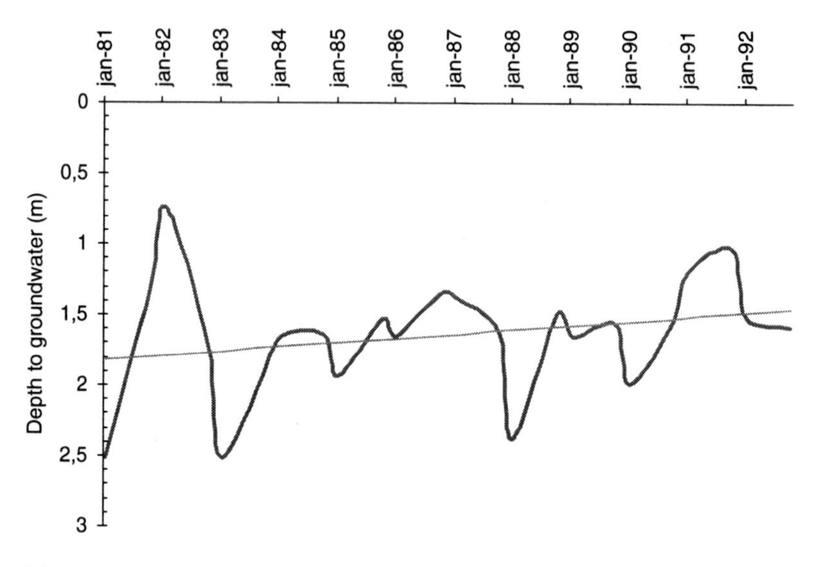

Fig. 4.6 Fluctuation of the depth to groundwater under an irrigated area

$$depleted\ fraction = \frac{ET_{a,gross}}{V_c + P} \tag{4.9}$$

Figure 4.7 shows the change in groundwater level (meter per month) for sandy loam soils in the Nilo Coelho irrigation command area, Brazil. Similar shaped relations can be established for other irrigated areas (Bos 2004). For semi-arid and arid regions the straight line in Fig. 4.7 intersects the x-axes at a depleted fraction between 0.5 and 0.7 (average about 0.6).

In other words: if $ET_{a,gross}$ is less than about $0.6(P + V_c)$ a portion of the available water goes into storage causing the groundwater table to rise while storage decreases if $ET_{a,gross}$ is greater than $0.6(P + V_c)$. Apparently, the natural drainage from command areas in arid and semi-arid regions has a capacity that is sufficient to discharge about $0.4(P + V_c)$. Thus, the depleted fraction can be used as a performance indicator on irrigation water use. The volume of water diverted into the irrigated area can be reduced during months with a low depleted fraction. If this non-diverted water remains in a storage reservoir (which often is the case in arid and semi-arid regions) this water can be diverted during warmer months.

As discussed in Section 1.3, $ET_{a,gross}$ is composed of three parts; ET_a from the irrigated (cropped) area, $ET_{a,fallow}$ from the irrigable non-cropped area, and the $ET_{a,non\text{-}ir}$ from the permanently non-irrigated part of the command area. Hence,

$$ET_{a,gross} = ET_a + ET_{a,fallow} + ET_{a,non\text{-}ir} \tag{4.10}$$

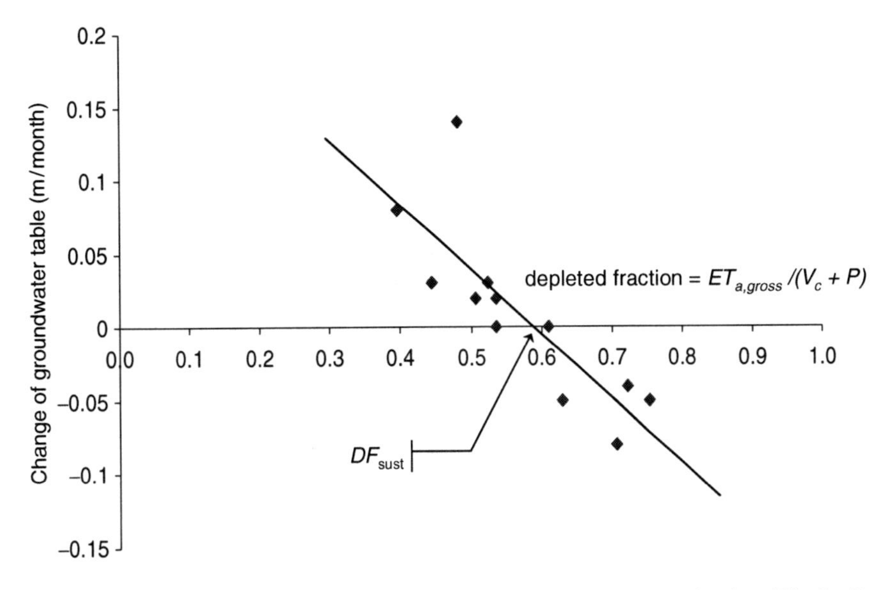

Fig. 4.7 Monthly change of groundwater table as a function of the depleted fraction, Nilo Coelho irrigated command area, Brazil (Bos 2004)

The parameters, V_c and $ET_{a,gross}$ in Equation 4.9 are not entirely independent of each other. As long as there is sufficient irrigation water, the ET_a-part in Equation 4.10 will be near its potential value. However, if V_c is reduced in order to increase the depleted fraction, less water will be available for irrigation and ET_a may decrease. This impact of V_c *on* ET_a is illustrated in Fig. 4.8 for the Fayoum depression, Egypt. As shown, the evaporative fraction, ET_a/ET_p for the irrigated area remains about unity if the depleted fraction is less than 0.6. During part of the year such a high evaporative fraction is needed in order to leach accumulated salts, etc. from the root zone of the crop. For higher values of the depleted fraction the value of ET_a/ET_p decreases by about 20%. Due to the shape of the yield versus *ET* curve of most crops (see Fig. 1.5), a decrease within this range results in a higher yield per cubic meter of water. However, crop yield per hectare will decrease. Therefore, in order to sustain agriculture on the one hand (leaching of the root zone is needed) and to attain a high productivity in terms of yield per cubic meter of water on the other hand, the monthly values of the depleted fraction should range between 0.4 and 0.9 while the annual average should be near the established intersection point for the irrigated area. The value of the depleted fraction at this intersection point between the straight line and the x-axes is the value for a *sustainable* groundwater balance. It is denoted as DF_{sust} (see Fig. 4.7).

The actual use of the depleted fraction as a performance indicator greatly depends on the method (and related cost) with which the parameters are quantified. Methods that provide sufficiently accurate data are summarized in Table 4.1.

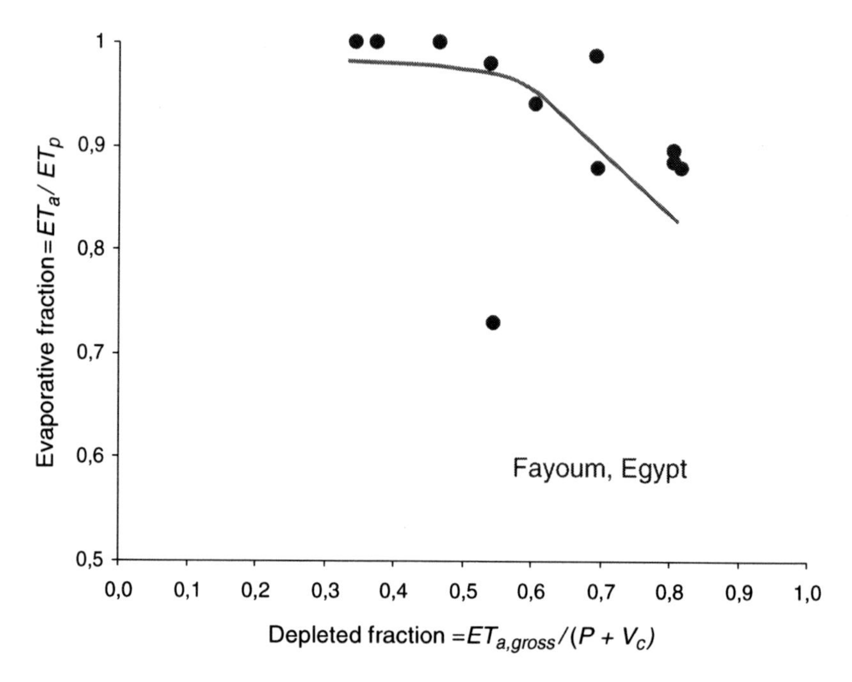

Fig. 4.8 Change in the relative ET of the irrigated area as a function of the depleted fraction of the gross command area, Fayoum, Egypt (Bos 2004)

4.5 Rooting Depth

If the water that rises above the groundwater table through capillary action reaches the effective rootzone of the irrigated crop, the crop may consume this water. Hence, in order to estimate the capillary contribution of water, we need information on the effective rooting depth of the crops in the area. For this we consider the four stages of crop development of Sections 1.2 and 2.10.

- During the initial growth stage, the crop just has been planted or seeded. For all crops we assume an initial rooting depth of 0.05 m. Thus, capillary water should virtually reach the soil surface in order to become a water source. For this capillary water not to cause salinity, a well-drained soil and sufficient off-season precipitation is needed.
- During crop development the above-ground biomass production grows in proportion with the root system. The depth to which the roots penetrate the soil firstly depend on the crop. Secondly, the root depth depends on the ease with which the soil can be penetrated (soil type and texture) and on the depth over which there is oxygen in the soil (most crops have no roots below the groundwater table) and on the need to search for water at greater depth (shallower roots with higher irrigation frequency). At the end of the development stage the effective root system can often be taken to be equal to that of a fully grown crop.

Table 4.1 Water management parameters and their method of measurement. Shown errors are for one single measurement

Parameter	Method by which term is measured or source of data
Actual evapo-transpiration, $ET_{a,gross}$ and ET_a	The spatial variation of $ET_{a,gross}$ can be calculated from the energy balance of the pixels of a satellite image having thermal bands (20% error). High resolution images (Landsat or Aster) can be used for detailed studies on ET_a. Low resolution images (NOAA or MODIS) are adequate for calculating values for areas greater than 2000 ha (Bandara 2006) and have the advantage of better temporal availability.
Potential evapo-transpiration, ET_p	As discussed in Chapter 2, ET_p can be calculated from a variety of equations. The most widely tested is Penman-Monteith (error 20%) (Allen et al. 2002; Burt et al. 2002).
Volume of water diverted from river, V_c	V_c should be measured with a permanent flow measurement structure. If the volume of water is calculated from 15 or more individual flow measurements (readings) the error in the volume of water will be reduced to the systematic error in these measurements (e.g. undershot gates 5%, broad-crested weirs 2%).
Precipitation, P	As discussed in Chapter 3, precipitation is measured with a gage that is installed in accordance to standardized rules (error 5%). Data commonly are already available from local meteorological stations. The spatial distribution of precipitation can be obtained from weather satellite data (error 10%).
Depth to ground water table	The groundwater depth is measured by lowering a sounder or installing a transducer into an observation well. The random error is about 0.02 m. A systematic error of 0.05 m can occur in the reference elevation of the ground surface.

- During the mid-season and late-season growth stages the effective rooting depth of the crop can be estimated from Table 4.2. The larger values are for soils having no significant layering or other characteristics that can restrict rooting depth. The smaller values for the rooting depth may be used for irrigation scheduling and the larger values for modeling soil water stress or for rainfed conditions.

4.6 How To Correct for Capillary Rise?

As discussed above, the balance between seepage and capillary rise is influenced by several parameters (precipitation, irrigation water delivery, ET_a, groundwater depth, rooting depth, capillary rise). To minimize the cost to quantify and handle these parameters for each irrigated field throughout the growing season, we recommend a step-wise pragmatic approach:

- The depth to the groundwater table. This can be measured in an observation tube, and because the groundwater table fluctuates gradually, only one measurement per week will usually suffice (Fig. 4.9).

Table 4.2 Typical crop height, ranges of maximum effective rooting depth for common crops (From FAO-56; Allen et al. 1998)

Crop	Maximum crop height (m)	Effective root depth (m)
a. Small Vegetables		
Broccoli	0.3	0.4–0.6
Brussel Sprouts	0.4	0.4–0.6
Cabbage	0.4	0.5–0.8
Carrots	0.3	0.5–1.0
Cauliflower	0.4	0.4–0.7
Celery	0.6	0.3–0.5
Garlic	0.3	0.3–0.5
Lettuce	0.3	0.3–0.5
Onions – dry	0.4	0.3–0.6
– green	0.3	0.3–0.6
– seed	0.5	0.3–0.6
Spinach	0.3	0.3–0.5
Radishes	0.3	0.3–0.5
b. Vegetables – Solanum Family (*Solanaceae*)		
Egg Plant	0.8	0.7–1.2
Sweet Peppers (bell)	0.7	0.5–1.0
Tomato	0.6	0.7–1.5
c. Vegetables – Cucumber Family (*Cucurbitaceae*)		
Cantaloupe	0.3	0.9–1.5
Cucumber – fresh market	0.3	0.7–1.2
– machine harvest	0.3	0.7–1.2
Pumpkin, Winter Squash	0.4	1.0–1.5
Squash, Zucchini	0.3	0.6–1.0
Sweet Melons	0.4	0.8–1.5
Watermelon	0.4	0.8–1.5
d. Roots and Tubers		
Beets, table	0.4	0.6–1.0
Cassava – year 1	1.0	0.5–0.8
– year 2	1.5	0.7–1.0
Parsnip	0.4	0.5–1.0
Potato	0.6	0.4–0.6
Sweet Potato	0.4	1.0–1.5
Turnip (and Rutabaga)	0.6	0.5–1.0
Sugar Beet	0.5	0.7–1.2
e. Legumes (*Leguminosae*)		
Beans, Green	0.4	0.5–0.7
Beans, Dry and Pulses	0.4	0.6–0.9
Beans, Lima, Large vines	0.4	0.8–1.2
Chick pea	0.4	0.6–1.0
Fababean (broad bean)		
– fresh	0.8	0.5–0.7
– dry/seed	0.8	0.5–0.7
Grabanzo	0.8	0.6–1.0
Green Gram and Cowpeas	0.4	0.6–1.0
Groundnut (Peanut)	0.4	0.5–1.0
Lentil	0.5	0.6–0.8
Peas – fresh	0.5	0.6–1.0
– dry/seed	0.5	0.6–1.0

(continued)

Table 4.2 (continued)

Crop	Maximum crop height (m)	Effective root depth (m)
Soybeans	0.5–1.0	0.6–1.3
f. Perennial Vegetables (with winter dormancy and initially bare or mulched soil)		
Artichokes	0.7	0.6–0.9
Asparagus	0.2–0.8	1.2–1.8
Mint	0.6–0.8	0.4–0.8
Strawberries	0.2	0.2–0.3
g. Fiber Crops		
Cotton	1.2–1.5	1.0–1.7
Flax	1.2	1.0–1.5
Sisal	1.5	0.5–1.0
h. Oil Crops		
Castorbean (*Ricinus*)	0.3	1.0–2.0
Rapeseed, Canola	0.6	1.0–1.5
Safflower	0.8	1.0–2.0
Sesame	1.0	1.0–1.5
Sunflower	2.0	0.8–1.5
i. Cereals		
Barley	1	1.0–1.5
Oats	1	1.0–1.5
Spring Wheat	1	1.0–1.5
Winter Wheat	1	1.5–1.8
Maize, Field (grain) *(field corn)*	2	1.0–1.7
Maize, Sweet *(sweet corn)*	1.5	0.8–1.2
Millet	1.5	1.0–2.0
Sorghum – grain	1–2	1.0–2.0
– sweet	2–4	1.0–2.0
Rice	1	0.5–1.0
j. Forages		
Alfalfa – for hay	0.7	1.0–2.0
– for seed	0.7	1.0–3.0
Bermuda – for hay	0.35	1.0–1.5
– Spring crop for seed	0.4	1.0–1.5
Clover hay, Berseem	0.6	0.6–0.9
Rye Grass hay	0.3	0.6–1.0
Sudan Grass hay (annual)	1.2	1.0–1.5
Grazing Pasture – rotated grazing	0.15–0.30	0.5–1.5
– extensive grazing	0.10	0.5–1.5
Turf grass – cool season[1]	0.10	0.5–1.0
– warm season[1]	0.10	0.5–1.0
k. Sugar Cane	3	1.2–2.0
l. Tropical Fruits and Trees		
Banana – 1st year	3	0.5–0.9
– 2nd year	4	0.5–0.9
Cacao	3	0.7–1.0
Coffee	2–3	0.9–1.5
Date Palms	8	1.5–2.5
Palm Trees	8	0.7–1.1
Pineapple	0.6–1.2	0.3–0.6
Rubber Trees	10	1.0–1.5

(continued)

Table 4.2 (continued)

Crop	Maximum crop height (m)	Effective root depth (m)
Tea – non-shaded	1.5	0.9–1.5
– shaded	2	0.9–1.5
m. Grapes and Berries		
Berries (bushes)	1.5	0.6–1.2
Grapes – table or raisin	2	1.0–2.0
– wine	1.5–2	1.0–2.0
Hops	5	1.0–1.2
n. Fruit Trees		
Almonds	5	1.0–2.0
Apples, Cherries, Pears	4	1.0–2.0
Apricots, Peaches, Stone Fruit	3	1.0–2.0
Avocado	3	0.5–1.0
Citrus		
– 70% canopy	4	1.2–1.5
– 50% canopy	3	1.1–1.5
– 20% canopy	2	0.8–1.1
Conifer Trees	10	1.0–1.5
Kiwi	3	0.7–1.3
Mango	5	1.5
Olives (40–60% ground coverage by canopy)	3–5	1.2–1.7
Pistachios	3–5	1.0–1.5
Walnut Orchard	4–5	1.7–2.4

[1]Cool season grass varieties include bluegrass, ryegrass and fescue. Warm season varieties include Bermuda grass, buffalo grass and St. Augustine grass. Grasses are variable in rooting depth. Some root below 1.2 m while others have shallow rooting depths. The deeper rooting depths for grasses represent conditions where careful water management is practiced with higher depletion between irrigations to encourage the deeper root exploration.

- The effective rooting depth of the irrigated crop. As discussed in Section 4.5, the actual rooting depth of a crop varies with the type, variety, and age of the crop, with the soil type and texture, with the depth to the groundwater table, with the irrigation frequency, and so on. As the actual rooting depth of irrigated crops is difficult to measure, only rough information will be available on this subject. Table 4.2 can be used for a preliminary estimate of the effective rooting depth of fully-grown crops (Fig. 4.9).
- For the common soils in the area, determine the height of capillary rise (C) for an upward flow rate of 0.5 and 2.0 mm/day (see Figs. 4.4 and 4.10).
- Subtract the effective rooting depth from the depth to the groundwater table. Check to see if this difference is less than the capillary rise (C) at a unit flow rate[1] of 0.5 and 2 mm/day (Fig. 4.10).
- If so estimate the capillary flow (Fig. 4.10).

[1]CRIWAR ignores a capillary flux of less than 0.5 mm/day because of the error in the estimate of this flux. The 2 mm/day value is selected because it represents a significant contribution of water.

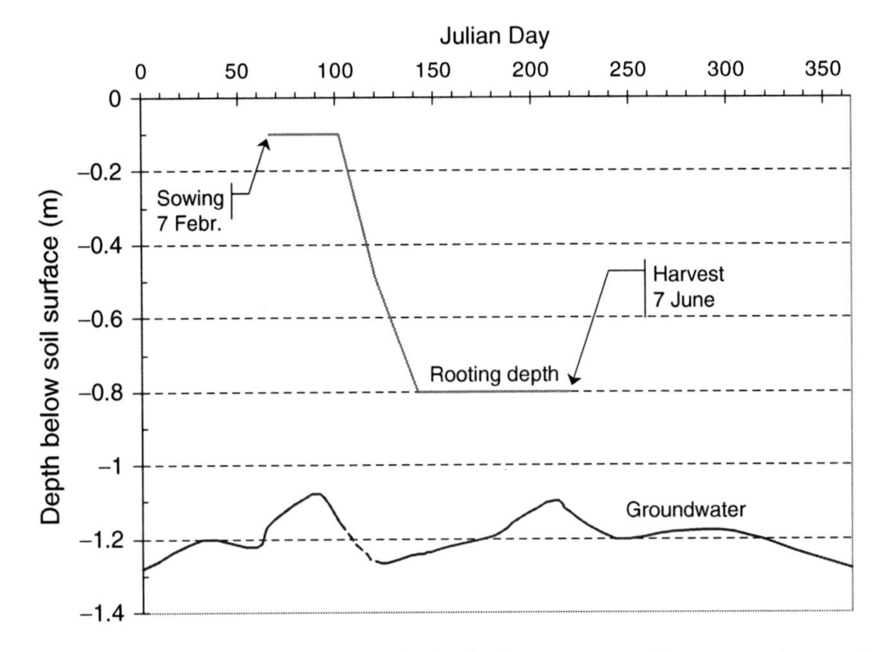

Fig. 4.9 Development of the effective rooting depth of a tomato crop with respect to the groundwater depth

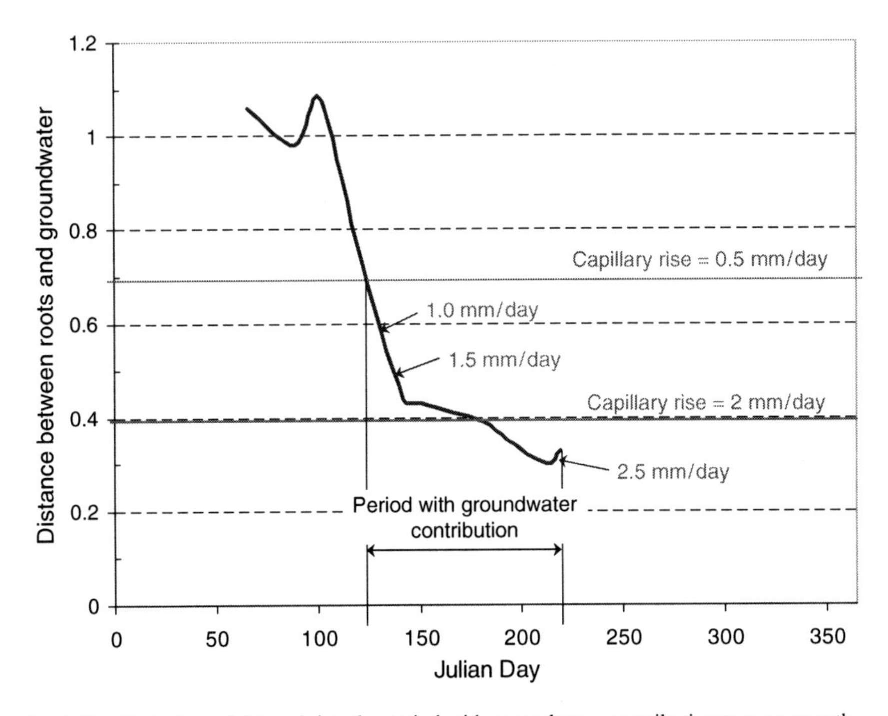

Fig. 4.10 Illustration of determining the period with groundwater contribution to crop growth

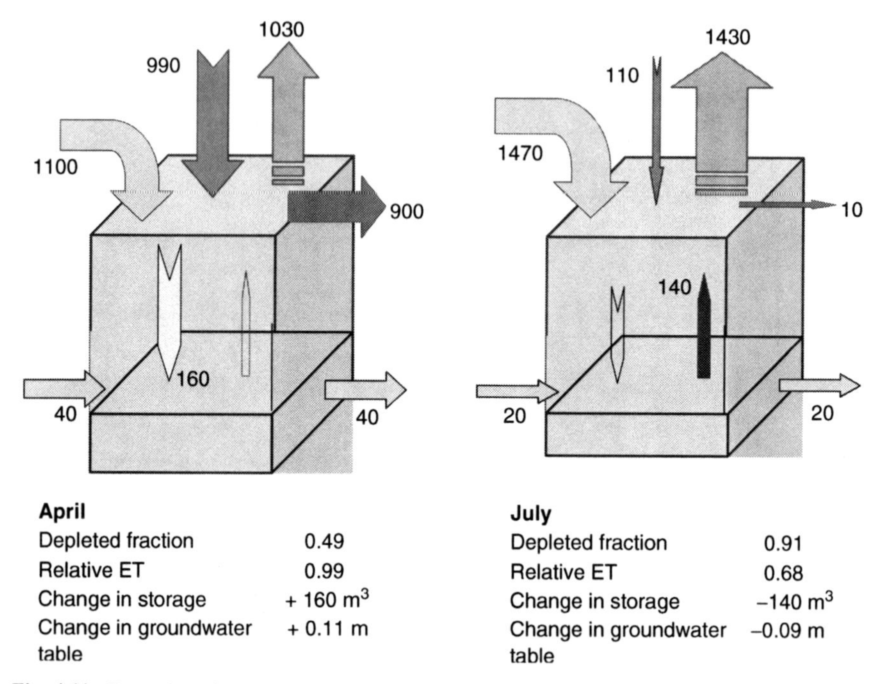

Fig. 4.11 Examples of a monthly water balance. All flows are in m^3/ha per month

The above capillary flow can be consumed by the crop and thus can be subtracted from the'crop irrigation water requirements' in order to calculate the volume of water that needs to be applied to the field(s) (also see Chapter 5).

This is illustrated in Fig. 4.11. In April the depleted fraction for an irrigation unit is low ($ET_{a,gross}/(P + V_c) = 0.49$) resulting in a relatively high ET_a from the irrigated fields (with $ET_a/ET_p = 0.99$). Part of the available water seeps through the root zone and causes the groundwater table to rise. This seepage is used to leach accumulated salts from the root zone. Thus, although capillary rise could meet part of the crop water requirements, the water manager does not use this option – he opts for groundwater storage and leaching. During water scarcity in July, the depleted fraction is high resulting in a lower (yet acceptable) relative ET_a. During this month water is taken from storage through the capillary rise. As a result, salts accumulate in the root zone.

In order to determine if capillary water is used as a water source during a certain part (month) of the irrigation season, the water manager needs to establish the relationship between the rate of change of the groundwater table and the depleted fraction (see Fig. 4.7). Using the intersection point as the annual average value of the depleted fraction (DF_{sust}) a monthly distribution of the depleted fraction should be decided upon. Selection of the distribution of monthly DF-values determines the degree to which capillary water is needed as a water source. CRIWAR uses this distribution in evaluating the water balance strategy of the irrigated command area.

Chapter 5
Irrigation Water Requirements

5.1 The Concept

The ultimate purpose of irrigated agriculture is to apply an 'intended' volume of water to crops in order to avoid undesirable stress throughout the growing cycle. The volumes of water flowing through a 'typical' irrigation (and drainage) system are illustrated in Fig. 5.1. As can be seen, the system depends on the upstream environment and influences the downstream environment. Within the system there is a strong interaction between surface water and groundwater and vice versa. In order to avoid the accumulation of salt etc. within the irrigated area, more water needs to enter the area ($P + V_c$) than is 'consumed' (i.e. $ET_{a,gross}$) by all crops and non-irrigated land. The non-consumed part of ($P + V_c$) returns to the groundwater basin or flows into the downstream surface water system. Provided that the quality of this return flow is acceptable, it can be re-used downstream. In many river basins, water is used and re-used by a variety of agricultural, environmental, urban, industrial, and recreational users. During this use and re-use, up to 90% of all water in the basin may be consumed before environmental degradation occurs in the downstream part of the basin. Thus, the environmental flow for river reaches and for downstream wetlands should be quantified more accurately as part of a water allocation plan.

Figure 5.1 shows the interaction of two sorts of 'water flows' that should be managed in such a way that crops can be grown in the command area:

- The classical flow of irrigation water from the surface water source (river diversion or reservoir) through the conveyance and distribution system to the fields
- The less visible (and often ignored) vertical flows of water seeping from the canals and fields to the groundwater basin and the 'return' flow through pumping and capillary rise

In order to calculate the volume of irrigation water, $V_{c,i}$ required to be delivered through the flow control structure serving the ith command area, we divide the crop irrigation water requirements, $ET_p - P_e$, by the relevant irrigation water use ratios. These ratios quantify the hydraulic functioning of components of the irrigation system in a spatial context over a specific time period. Depending on the command

M.G. Bos et al. *Water Requirements for Irrigation and the Environment*,

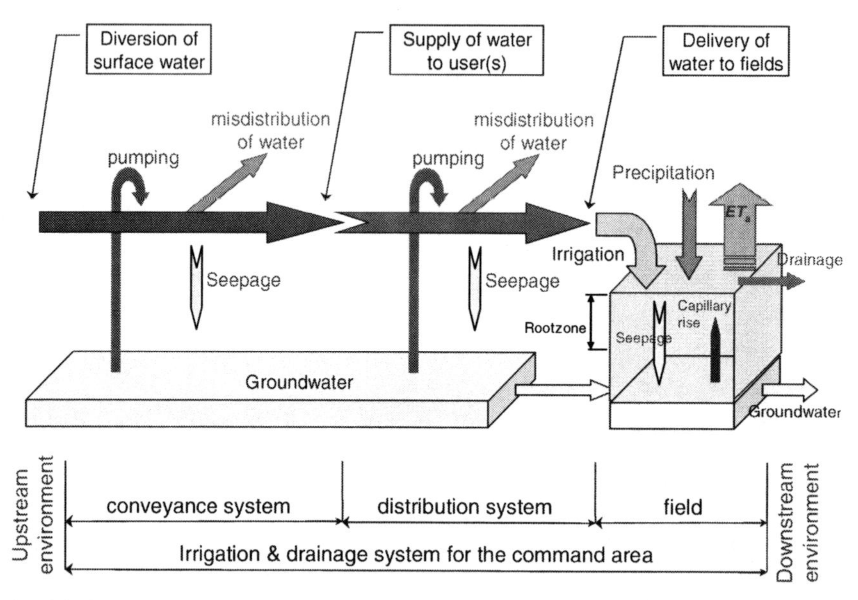

Fig. 5.1 Schematic water flows in an irrigation and drainage system

area under consideration, the relevant ratios are combinations of the field application ratio; the distribution ratio and the conveyance ratio. These ratios will be discussed in Sections 5.2–5.4.

5.2 The Field

Looking back over several thousand years, irrigators have developed a wide variety of methods in order to apply water to a field. All methods were designed to apply water as uniformly as possible to all plants so that water stress is limited. Depending on the used level of technology, each method has the ability to apply water with a related uniformity. However, all methods apply more water to some plants in a field and less to others. Because farmers tend to apply sufficient water to the driest part of the field, most of the field gets more water than required. The volume of irrigation water that is (or needs to be) delivered at the field inlet thus depends on the value of $(ET_p - P_e)$ of the irrigated crop and on the uniformity with which water can be applied. Water need and water delivery are related to each other through the field application ratio, R_a. The ICID standard definition for the field application ratio (efficiency) is (Bos 1997)

$$R_a = \frac{V_m}{V_f} \quad 5.1$$

Where,
V_m = volume of irrigation water needed, and made available, to avoid undesirable stress in the crops throughout the growing cycle (m^3/period)
V_f = volume of irrigation water delivered to the fields during the period under consideration (m^3/period)

The value of V_m in Equation 5.1 is difficult to establish on a real-time basis because many complicated field measurements would be needed. However, the method that is used to quantify V_m is not very important, provided that the same (realistic) method is used for all command areas (lateral or tertiary units) within the

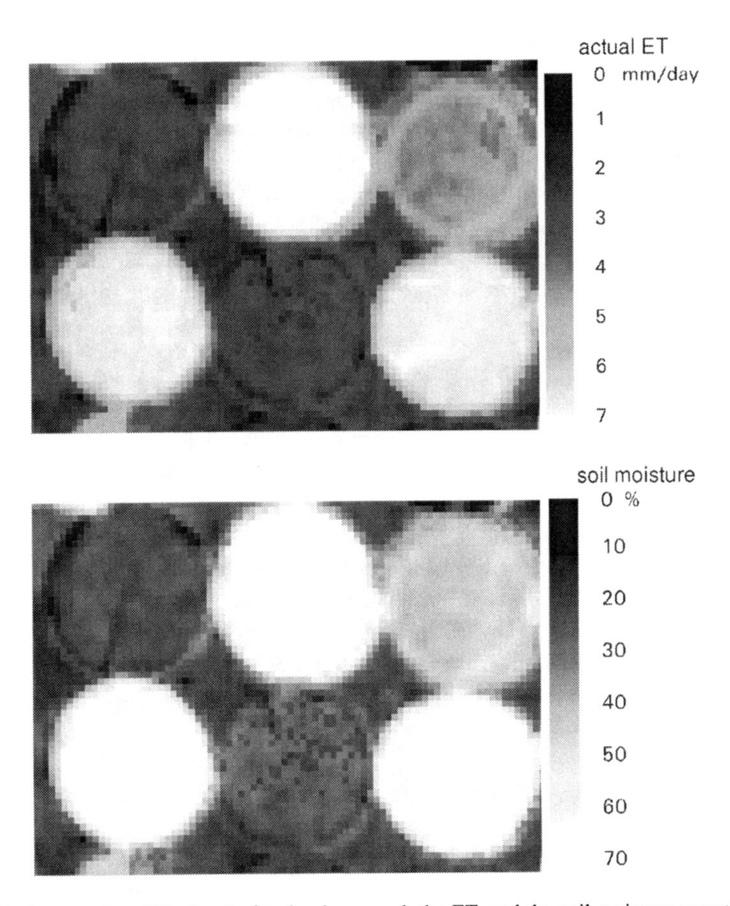

Fig. 5.2 Irrespective of the level of technology used, the ET_a and the soil moisture vary throughout the irrigated field. The images above show soil/moisture and ET_a from center pivots, Washinton State, USA, Landsat 25 July, 2000. (courtesy WaterWatch, Wageningen).

irrigated area. For practical purposes, we can assume that V_m equals the evapotranspiration by the irrigated crop, minus the effective part of the precipitation (i.e. the $ET_p - P_e$ as calculated by CRIWAR). The irrigation water requirement at the field inlet then equals:

$$V_f = \frac{ET_p - P_e}{R_{a,target}} \quad 5.2$$

The target value of the field application ratio, $R_{a,target}$, depends on the level of technology used to apply water, on the level of aridity of the climate, on the availability of irrigation water, and on crop characteristics (dry-foot crop or ponded rice). How they can be determined is shown below.

5.2.1 Dry-Foot Crops

The ability of an irrigation technology to apply water uniformly to a field is an important criterion in determining the level of technology to be used. At the same time, this uniformity influences the volume of water (per irrigation turn) that needs to be applied to the field, in addition to the crop irrigation water requirements. As an example, let us consider a level basin to which $V_m = 100$ mm needs to be applied for the considered turn (Fig. 5.3). If the actually applied water depths, $V_{a,i}$, (applied volume or depth per irrigation turn) to parts of an irrigated field are measured, we can assume:

$$V_f = \sum V_{a,i} \quad 5.3$$

If the irrigator would decide to apply a volume V_f to the field being exactly equal to V_m, the field application ratio is 1.0 (100% efficiency). Nevertheless, 50% of the field has then been given more water than V_m; the other 50% has received less. In the part of the field that has received less, the ET_a will be less than ET_p and as a result, salt may accumulate in the root zone. This would not cause a problem if sufficient off-season precipitation is available to leach these salts. Hence, the fraction, F, of the field that is allowed to receive less water than $V_m = ET_p - P_e$ depends on the climate.

Till and Bos (1985) assumed a normal distribution of $V_{a,i}$ and recommended that the summed target flow to avoid water stress and salt accumulation to a field (or volume of flow over a considered period) equals

$$V_{f,target} = \left(\sum V_{a,i}\right)_{target} = \left(1 + sT_p\right) \times \sum V_{m,intended} \quad 5.4$$

Where, the standard deviation, s, of the water application ratio, $V_{a,i}/V_f$, should be measured for an applied volume (or depth) of water that approximates $V_{m,intended}$. The latter depends on the depth of water applied due to the uniformity of the water application. For the example of Fig. 5.3, the value of s equals 0.11.

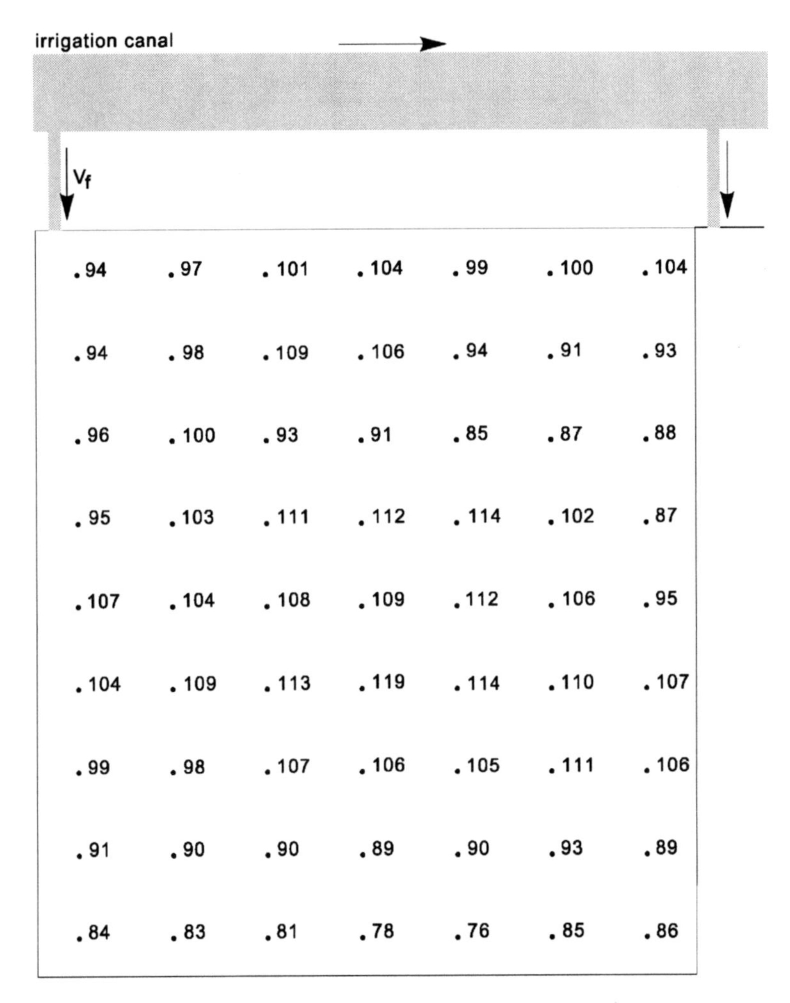

Fig. 5.3 Measured depths (V_{ail} in mm/turn) of irrigation water applied to a level basin (Till and Bos 1985)

T_p is a statistical value that is exceeded by a random variable, normally distributed, with zero mean, and with standard deviation units. Values of T_p versus F are listed in statistical handbooks. An extract is given in Table 5.1.

As shown above, the target value of V_f depends on the standard deviation, s, of the 'irrigation water application' and on the fraction of the field where a water shortage is acceptable (F in %). The standard deviation depends on the level of technology available to apply water uniformly and on the 'quality of management and on operation by the farmer'. As mentioned earlier, the percentage of the area where a water shortage is acceptable depends on the climate. Till and Bos (1985) recommend a T_p-value of 0.67 (F is about 25%) if off-season precipitation is available to leach the accumulated salts. In arid and semi-arid climates, this precipitation

Table 5.1 Values of T_p versus F

F (in %)	T_p (dimensionless)
50	0
25	0.67
10	1.28
5	1.64
2.5	1.96
1.0	2.33

may not be available. Then a value of $T_p = 2.0$ (F is about 2.5%) is recommended. The target value of the field application ratio for dry-foot (non-rice) crops is then,

$$R_{a,target} = \frac{V_m}{(1+sT_p)\times V_m} \tag{5.5}$$

Figure 5.4 shows values of $R_{a,target}$ as a function of the level of technology (the standard deviation of water application) and the part of the field that may receive less than the intended water need (F in percent of field).

If the field of Fig. 5.3 is in a climate with sufficient rain to leach accumulated salts ($F = 25\%$), Equation 5.5 gives:

$$R_{a,target,humid} = \frac{100}{(1+0.11\times 0.67)\times 100} = 0.93$$

In arid climates, the fraction F should be as low as 2.5%. Hence,

$$R_{a,target,arid} = \frac{100}{(1+0.11\times 2.00)\times 100} = 0.82$$

Substitution of the latter two target values into Equation 5.2 shows that, under arid conditions, the required volume of irrigation water, V_f, is 0.93/0.82 = 1.13 times greater than under more humid conditions. This extra water is needed for sustainable agriculture. Since water is a scarce resource in arid zones, its efficient use would require a higher level of technology and related management (smaller value of s). As shown in Fig. 5.4, the standard deviation of water application needs to be better than 0.17 in order to enable an acceptable target value of R_a.

5.2.2 *Paddy Rice*

For paddy rice, the ICID (Senga and Mistry 1989) recommended that the seepage from the field, $V_{f,seepage}$, be added to the target volume of water application. Hence,

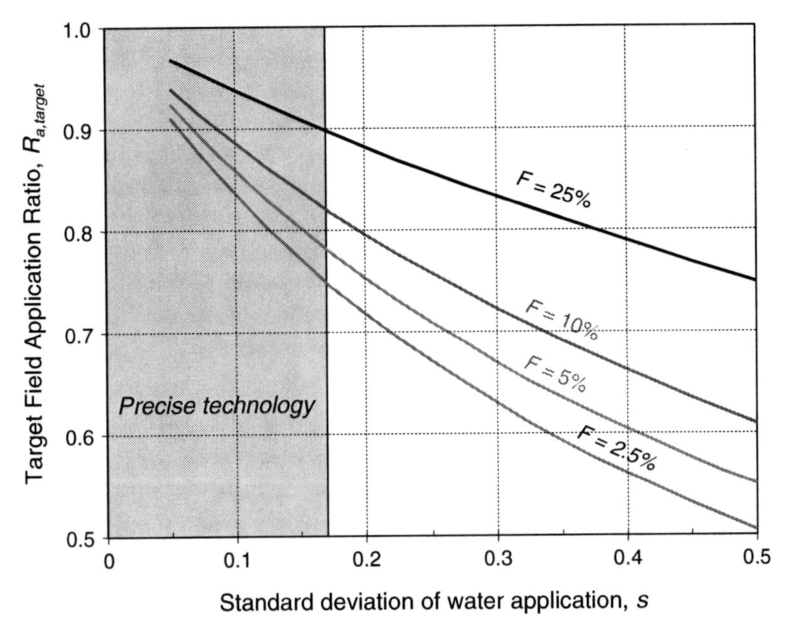

Fig. 5.4 Values of $R_{a,target}$ as a function of the level of technology (the standard deviation of water application) and the part of the field that may receive less than the intended water need (F in percent of field)

$$R_{a,\ target\ paddy} = \frac{V_m}{(1 + sT_p) \times V_m + V_{f,seepage}} \quad 5.6$$

For well-levelled fields with ponded water, the values of both s and T_p approach zero. Equation 5.6 shows that the target ratio for paddy rice decreases with increasing seepage from the field. A lower limit should be set to the target field application ratio; if there is too much seepage, the paddy should not be grown.

5.2.3 Water Application Methods

In order to illustrate the above relationship between crop production, uniformity of water application (and the related field application ratios), several irrigation methods are discussed below.

Furrows, laser graded. This is the highest level of technology available with furrow irrigation. In combination with skilled flow control a reasonably high uniformity of water application is possible ($s \cong 0.25$).

Furrows, other quality grading. Low quality grading makes it difficult for the operator (farmer) to apply sufficient water to all parts of the field. Together with poor flow control this often leads to low uniformity of water application ($s >> 0.5$). The field application ratio often is less than 40%. A poorly graded furrow is difficult to operate and is the least efficient water application method.

Border strips, laser graded. From a hydraulic and water management point of view, border strips are 'wide furrows'. Because of this width, the flow rate per strip is proportionally greater. The operator (farmer) needs to be careful that the bund at the downstream end of the strip does not break. In combination with skilled flow control a reasonably high uniformity of water application is possible ($s \cong 0.25$).

Border strips, other quality grading. Because of its width, flow in a border strip is sensitive to cross-slope (perpendicular to the flow direction. Bunds are used to direct water over the full width of the strip. However, because of the cross-slope, uniformity will be lower than above (here about $s \cong 0.3$).

Level basin. Laser levelling allows a variation in land surface of about 1 cm. In this basin ridges were made to grow a row-crop (cotton). Water enters in between the ridges simultaneously and from both sides. With the proper matching of basin size, soil type and measured flow very high uniformities can be reached ($s \leq 0.1$). Thus, laser levelled basins allow very efficient water use (90%).

Level basin, Traditional levelling of basins often results in a wide variety of water depth on the field. If the flow rate into the basin is low (often the case with traditional basins) this results to a major difference between the 'opportunity time' for water to infiltrate in the lowest and highest part of the basin. Values of $s >> 0.5$ are common, resulting in inefficient water use.

Level basin, paddy rice. With well-levelled basins the value of $s \approx 0$. Thus, the only part of the applied water that is not consumed (ET_a) is the seepage (and drainage) from the field. This 'drainage water' may cause downstream waterlogging or pollution. In that case, it is recommended to set a limit on the percentage of applied water that is drained (e.g. 20%).

Sprinkler, hand-move system. Following water application, the 'first generation' sprinkler systems were moved to the next location for irrigation of the next strip of land. Because of problems with the nozzle alignment the spray pattern was fairly often non-circular. Variation in nozzle spacing also caused non-uniform water application. The value of s is rather high (>0.4) resulting to efficiencies of 60% or less.

Sprinkler, overhead rain drops. Irrigation machines (e.g. centre pivots and lateral move) were developed in order to save labour and water. Because of improved nozzle alignment, nozzle spacing and timing of water application, the uniformity improved considerably ($s \approx 0.25$). Target field application ratios of 70–75% are common for overhead 'rain drops'.

Sprinkler, downward fine spray. Multiple downward spraying nozzles reduce evaporation from 'rain drops' and increase uniformity ($s < 0.1$). These irrigation machines thus should (and can) operate at field application ratios (efficiencies) between 0.90% and 0.95%.

Drip irrigation differs from all other application methods because it applies water to the part of the field where a crop grows. As a result, salts accumulate at the wetting front. Provided that emitter clogging can be prevented (clean and filtered water is used) a value of $s \approx 0.10$ can be reached. Field application ratios as high as 90% can be targeted provided that off-season (winter) rain is available to leach accumulated salts.

Micro sprinkler partly uses the same technology as drip, except that the emitter is replaced by a small sprinkler. Because of the relative size of the hole through which water is applied, the sprinkler is less vulnerable to clogging. Also the wetted area is larger so that this method can be used to leach accumulated salts. The uniformity is slightly better than with drip ($s < 0.10$) so that water can be used efficiently (better than 90%).

5.3 The Distribution System

As shown in Fig. 5.1, the distribution system is the part of the irrigation system in between the conveyance system and the field inlet. Its function is to distribute the supplied water in accordance with an intended schedule to all water users within the related command area. In larger systems the distribution system usually coincides with the irrigation unit managed by the water users association. The distribution system then receives water from the irrigation district conveyance system and delivers water to individual farms. If farms are large (more than 50 ha), a single water user (farmer) usually receives water directly from the irrigation district and the distribution system then coincides with the on-farm system. Small systems (less than 100 ha) often have no conveyance system. The distribution system is then the entire system in between the water source and the farm/field inlet.

Due to the limited length of the distribution system, seepage often is less than 5% of the total flow (Bos and Nugteren 1974). However, delivering the intended flow rate at the intended time during an agreed period to a water user is complicated. Therefore, the misallocation of water (unaccounted delivery) represents a significant volume of water (Fig. 5.5). Besides the influence of the level of technology and the skills of the operator on the volume of misdistributed water, this volume is heavily influenced by social and economic boundary conditions.

As shown in Figs. 5.1 and 5.6, part of the irrigation water is diverted into the irrigated area while another part may be pumped from the aquifer. Pumping water into the distribution system has several advantages:

- The irrigation (canal) flow rate can be increased to meet short-term changes in water demand. This particularly improves the flexibility of operation for long travel times of water from its diversion point (on large systems).
- The groundwater table can be controlled so that water logging and salinity problems can be avoided. Non-consumed irrigation water that recharged the aquifer can be re-used so that less drainage water is discharged into the downstream environment.
- The aquifer can be used as a storage reservoir and supply water to all water users being served by the distribution canal during dry periods. Water seeping from

Fig. 5.5 In the distribution system the misdistribution of water usually exceeds seepage

Fig. 5.6 Groundwater being pumped into a distribution canal

canals and fields during wet months (low depleted fraction) can be re-used during dry months (high depleted fraction).

As mentioned above, the operational performance of the distribution system is influenced by the level of technology, by skills used by the system operator and by the social and political boundary conditions within which the operator performs. The quality of operation can be quantified by the *delivery performance ratio* Clemmens and Bos 1990; Molden and Gates 1990; Bos et al. 2005). In its basic form it is defined as:

$$\textit{Delivery Performance Ratio} = \frac{\textit{Actual Flow of Water}}{\textit{Intended Flow of Water}} \qquad 5.7$$

Depending on the availability of data the 'flow of water' can be determined in two ways (Fig. 5.7):

- In systems where no structures are available to measure the flow rate, time is the only remaining parameter to quantify water delivery performance. As shown in Fig. 5.7, the *Delivery Performance Ratio* then compares the actual length of the water delivery period with intended period. For operational purposes it then is assumed that the flow rate is constant during a relatively long period.
- For modern water management the flow rate must be measured (e.g. in m^3/s). Delivery performance of water then relates the actual delivered volume of water with respect to the intended volume. The length of the period for which the volume is calculated depends on the process that needs to be assessed. It varies

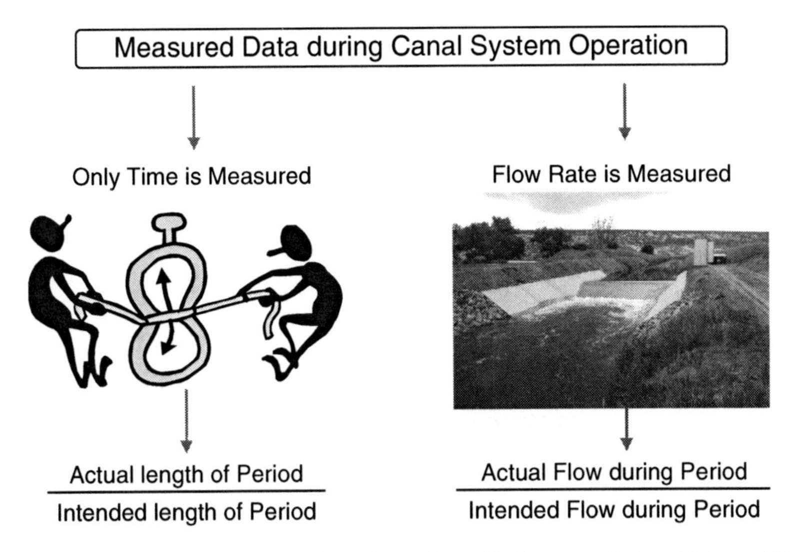

Fig. 5.7 Depending on the available data, the Delivery Performance Ratio will have different formats

from 1 s (for flow rate), one irrigation rotation interval (for water availability) to 1 month or year (for water balance) studies.

The *Delivery Performance Ratio* enables a water manager to determine the extent to which water is **actually** delivered as **intended** during a selected period and at any location in the system. It is obvious that if the actually delivered volume of water is based on frequent flow measurements, the greater the likelihood is that managers can match **actual** to **intended** flows. However, in order to obtain sufficiently accurate flow data, discharge measurement structures with water level recorders must be available at key water delivery locations (Bos 1976), and in order to facilitate the handling of data, recorders that write data on a chip are recommended (Clemmens et al. 2001).

The volume of water that needs to be supplied from the conveyance system into the delivery system is the sum of three components; the water that needs to be delivered to the fields $\left(\sum V_{f,i,intended}\right)$, the water needed as a sort of buffer to satisfy 'most' water users and water that seeps from the distribution system ($V_{d,seepage}$). The degree of satisfaction of the water users depends on the variation of water delivery in time and over the irrigated area.

The degree of satisfaction 'in time' can be re-phrased to the question: if the water user receives too little water during this irrigation turn, will this happen again during the next turn or will the user receive the intended share of the water? The variation can be documented by plotting the *delivery performance ratio* in time with respect to its target level that is based on the 'service agreement'. Around the target level is an allowable range (either to one or two sides) within which the indicator can fluctuate without triggering a complaint form the user or a management action. However, if the *delivery performance ratio* moves out of this range, diagnosis of the problem should lead to the planning of corrective action.

With respect to its spatial distribution, the indicator values of different irrigation units (lateral, tertiary or farms) within the same irrigated area can be compared and correlated with other parameters. Based on this information the water user can relate 'the received level of service' to that of neighbouring users.

Example

As an example of the use of the *delivery performance* ratio we consider the Los Sauces canal serving eighteen farms with a total area (with water rights) of 107.4 ha (Fig. 5.8). The unlined canal (Ramo Sauce) distributes water with the on-off system. In other words – all water entering the canal is delivered during a pre-announced period to one farm. In this irrigation scheme, the period of water delivery is related to the 'paid water rights' of the farmer. Thus, water is delivered during a pre-set period per farm. Random measurements have shown that the actual duration of flow into the farm inlets is almost equal to the intended delivery period. Gate opening & closure always is closely attended by two or three persons; the gate-man of the Users Association, the irrigator going to receive water, and (commonly) the irrigator who is going to end his turn. Because of this high dependability of water delivery we may use the intended delivery time in order to calculate the actual volumes of delivered water.

The intended water delivery in Table 5.2 is based on the measured flow at the intake structure of the Montecaseros Users Association (serving 8,581 ha). The UA intends to deliver this water in equal shares to all water users. The actual delivered flow is based on measured flow at the head of Los Sauce canal minus seepage in between this weir and the farm inlet. Average seepage in the Los Sauce canal is measured by subtracting the measured flow into the Blanco farm from the flow at the head of the canal. Seepage averages 1.25% per km of the inflow or about 1,5 m^3/turn. Similar data are available for other irrigation turns since the 1994/95 irrigation season.

The target volume of water per period (e.g. m^3/turn) that should be delivered to a group of water users within one irrigation unit depends on:

- The value of the standard deviation (s) of the *WDR* shown at the bottom row of Table 5.3.
- The interpretation of the concept 'agreed level of service' between the Users Association and the farmers. In this context this boils down to the volume of water that is intended to be delivery by the UA to the farmers within the irrigation unit (served through one control structure). In other words, which part (F in %) of the farms may receive less water than they need to meet all water requirements?
- The seepage from the considered water distribution system, $V_{d,seepage}$.

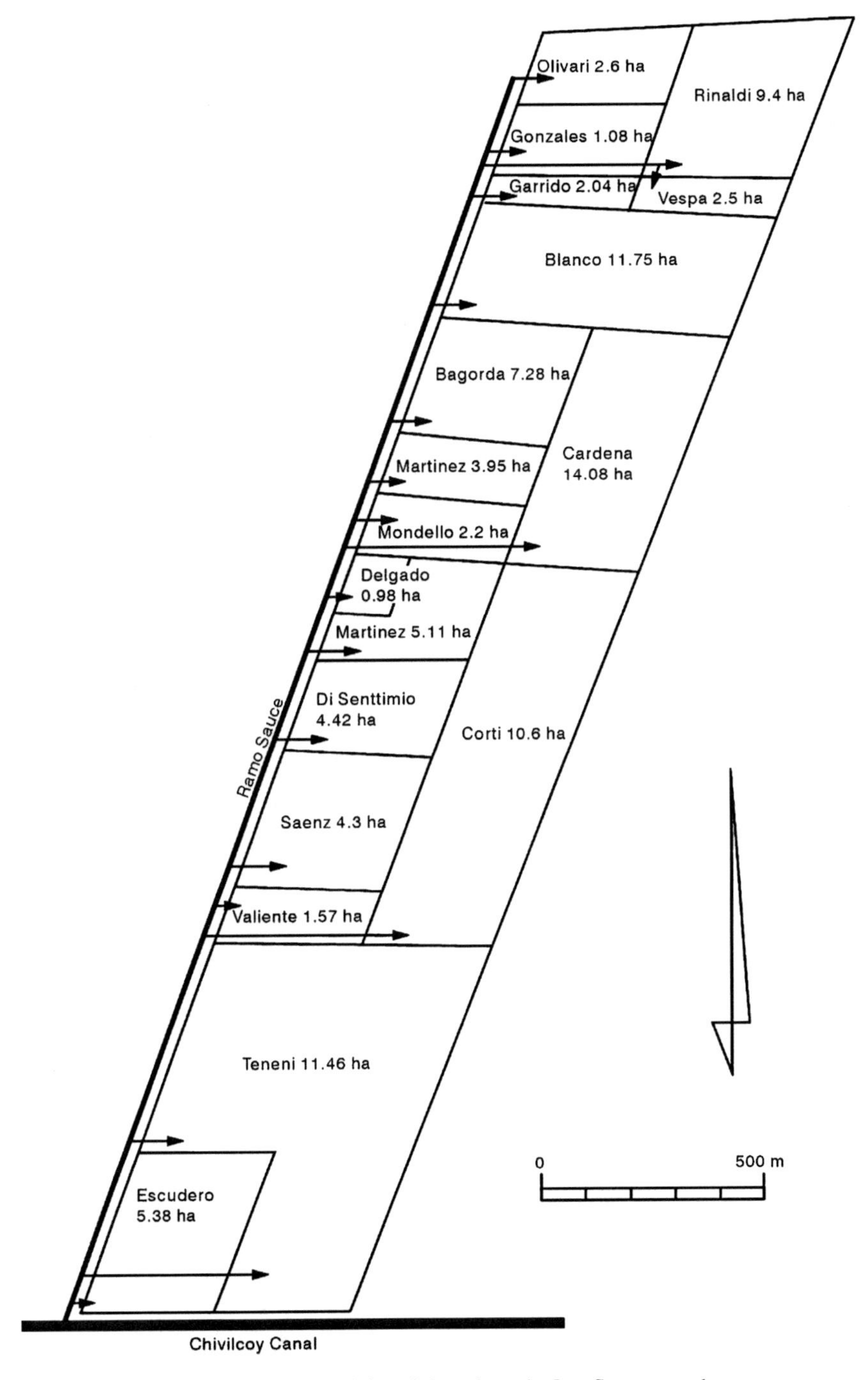

Fig. 5.8 Location of farms and related farm inlets along the Los Sauces canal

Table 5.2 Example of Water Delivery along the Los Sauces canal for the irrigation turn starting 3 January 1996 (Bos et al. 2001)

Name of water user	Area with water rights (ha)	Measured flow at Los Sauces (m^3/s)	Actually delivered volume (m^3/turn)	Intended volume delivered (m^3/turn)	Water delivery ratio (dimensionless)
Escudero	6	0,67	3,647	3,135	0.86
Terreni	12	0.77	7,295	7,169	0.98
Corti	12	0.65	7,295	5,992	0.82
Valiente	2	0.81	1,184	1,249	1.05
Saez	5	0.86	2,961	3,306	1.12
Disentimio	5	0.86	2,961	3,297	1.11
Martinez	7	0.84	4,147	4,496	1.08
Cardenas	18	0.79	10,662	10,795	1.01
Delgado	1	0.74	5,926	5,650	0.95
Mondello	4	0.74	2,370	2,256	0.95
Martinez	5	0.72	2,961	2,740	0.93
Bagorda	8	0.74	4,738	4,498	0.95
Blanco	14	0.72	8,293	7.631	0.92
Garrido	2	0.74	1,184	1,116	0.94
Vespa	3	0.72	1,777	1,616	0.91
Rinaldi	12	0.77	7,108	6,924	0.97
Gonzales	6	0.79	3,554	3,572	1.01
Olivari	5	0.79	2,961	2,971	1.00
Total =	127		77,379	78,413	
				Average	0.98
			standard deviation		0.08

Taking the aforementioned points into account, the summed target flow (or volume of flow over a considered period) serving a group of farm off-takes equals:

$$V_{d,target} = (1 + s\,T_p) \times \sum V_{f,i,intended} + V_{d,seepage} \qquad 5.8$$

The standard deviation, s, and the intended flow to be delivered to the i-th off-take, $V_{f,i,intended}$, were defined above. T_p is a statistical value that is exceeded by a random variable, F, normally distributed, with zero mean, and with standard deviation units. Values of T_p versus F are listed in Table 5.1.

As a first example, let us consider the measured water delivery during the turn of Table 5.2. If we assume that it is acceptable that 25% of the farmers receive less water than intended, the T_p value then is 0.67. Substituting the values of Table 5.2 into this equation yields for the considered irrigation turn:

$$V_{d,target} = (1 + 0.08 \times 0.67) \times 78413 + 1747 = 84360 \text{ m}^3 / \text{turn}$$

This volume of water exceeds the volume intended to be delivered by the Users Association for two reasons:

- Because the farmers receiving too little water differ per turn (no systematic under delivery of water), the farmers accept a percentage of 50% so that a value of T_p of zero can be used.
- In setting the intended water delivery, the UA ignores seepage between the head inlet structure of the canal system managed by the Montecaseros UA (4,800 ha) and the Los Sauces inlet. The UA also assumes that there is no misallocation of water between the tertiary units along the Montecaseros canal.

For the irrigation turn starting in 3 January 1996, the average *Delivery Performance Ratio* was $DPR = 0.98$. Hence, the average farmer received 2% less water than the UA intended to deliver on the basis of inflow at the inlet structure at the head of the Montecaseros canal.

The effectiveness (uniformity) of water delivery can be quantified by the standard deviation of the *delivery performance ratio*; hence, by the standard deviation of the measured $V_{f,actual}/V_{f,intended}$ values to outlets in the considered command area (irrigation unit). As shown in Table 5.2, the $s_{DPR} = 0.08$. Values below 0.1 indicate a very uniform water delivery. This high uniformity is attained by the gate operation practice. Gate opening and closure is accurately timed by two (or three) stakeholders; the gate operator, the irrigator who is going to receive water and (often) the irrigator who is about to end his turn. In order to reduce the standard deviation below 0.08, the inflow at the head of the canal should become as constant as practically possible.

As shown above, the target flow is strongly influenced by the standard deviation, s, of the delivery performance ratio and by the acceptability of water shortage (F in %). In calculating the above target volume, we tentatively assumed $F = 25\%$. In deciding on the proper F-value, irrigation managers will likely have to rely on past experience to see what an acceptable value is. The Los Sauces Unit receives water for 25 h once every 13 days. During the example turn of Table 5.2, the user that was furthest downstream received water first; during the next turn, he is last in line. As a result, the *DPR* for each user varies per turn and an F-value of 50% is accepted by the Los Cause farmers.

5.4 The Conveyance System

As discussed in Section 5.1, the conveyance system is the part of the irrigation system in between the water source (river diversion or reservoir) and the (group of) water users to which water is supplied. Water also can be pumped into the conveyance system from the local aquifer (groundwater basin). Basically there are three ways for water to leave the conveyance system (Fig. 5.9):

- Water being supplied to the distribution system. For modern water management these flows must be measured and controlled. Structures to be used for this purpose are described by Clemmens et al. (2001) and by the USBR (1997).
- Seepage through the bottom and sides of the canal system and leakage along gate seals. For a well maintained system this volume of water increases with the size of the system.

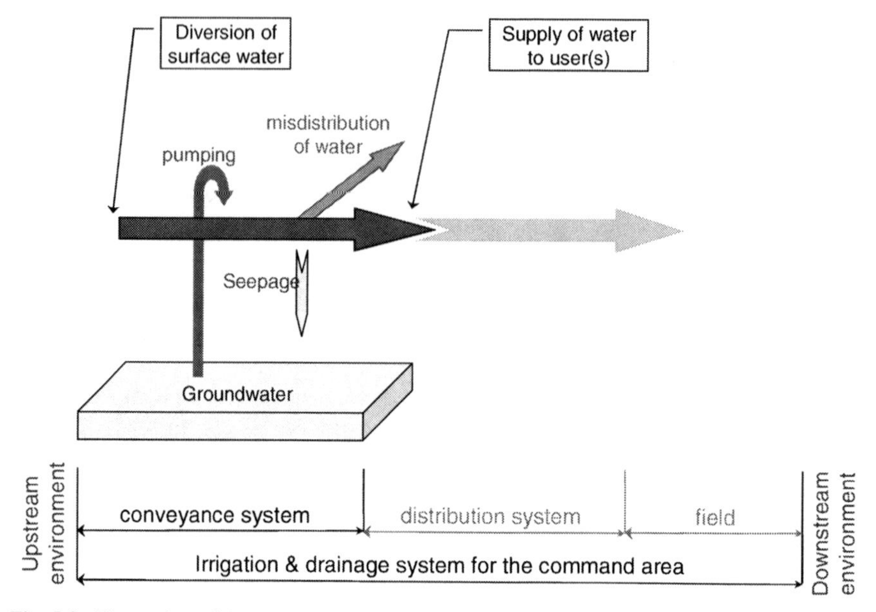

Fig. 5.9 The setting of the conveyance system

- The misallocation of water. This volume includes unaccountable supplies and the theft o f water.

The classical ratio used to quantify the water balance of a conveyance system (or reach of this system) is the *'outflow over inflow ratio'* (often named efficiency). The ratio has the structure:

$$Outflow\ over\ Inflow\ Ratio = \frac{Total\ Water\ Supply\ from\ Canal}{Total\ Water\ Diverted\ or\ Pumped\ into\ Canal} \qquad 5.9$$

The name commonly used for the ratio depends on the part of the system that is assessed. For large irrigation systems it is common to consider the conveyance ratio of parts of the system. In this chapter we consider the conveyance ratio of the upstream part of the system as managed by the Irrigation Authority. In mathematical terms the conveyance ratio reads:

$$Conveyance\ Ratio = \frac{V_d + V_{non-ir}}{V_c + V_{grw}} \qquad 5.10$$

Where:

V_d = total volume of irrigation water supplied to the inlets of the distribution system.

V_{non-ir} = total volume of water supplied for non-irrigation purposes. In most irrigation systems this volume is negligible with respect to V_d.

V_c = total volume of surface water diverted from the water source (river, reservoir) into the irrigation system.

V_{grw} = total volume of groundwater pumped into the conveyance system (Fig. 5.10).

The conveyance ratio should be calculated over a short (month) and a long (season) period (Fig. 5.11). Quantifying the outflow over inflow ratio for only 1 month gives information to the system manager provided that a target value of the ratio is known. A regular repetition of the measurement allows for an assessment of the trend, and gives an indication of time. This assists the manager in identifying processes that need to be reversed before remedial measures become too expensive or too complex. For example, a gradual decrease of the conveyance ratio means that seepage and mis-distributed water from the system increases. A diagnostic survey then can identify the cause (e.g. damaged lining, unauthorized off-take, leaking gate, malfunctioning water measuring device, etc.). The cause can then be removed as part of regular system management.

As shown in Fig. 5.9, seepage and misdistribution of water are major 'unintended' flows from an irrigation conveyance system. Seepage depends on the design and quality ('water tightness') of the system and on the area served (size) of the system. Figure 5.12 shows this relationship for canals with continuous flow.

For systems where canals are operated with fluctuating flow rates, or where canals are operated under a rotational delivery schedule, a significant volume of water can become 'unaccounted' for. The related misallocation of water (management losses in Fig. 5.12) reduces the conveyance ratio for small and large systems.

Fig. 5.10 Pumping groundwater into the (downstream) part of the conveyance systems improves the operational flexibility and control of the groundwater table

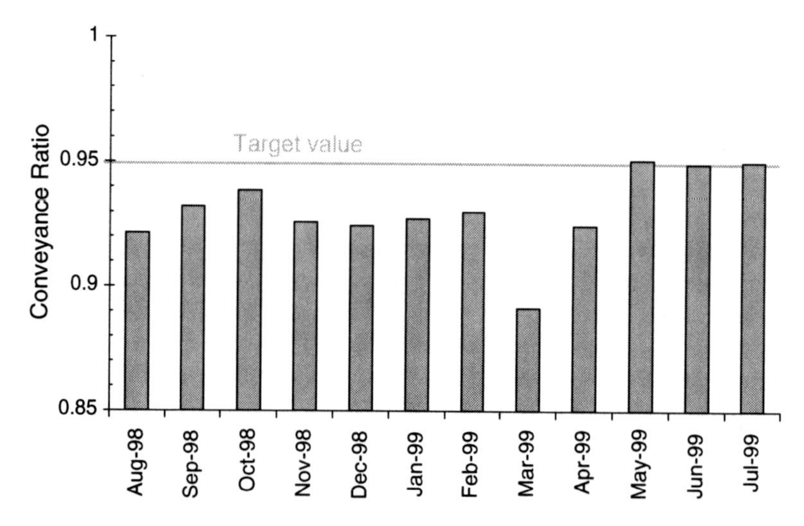

Fig. 5.11 Monthly values of the conveyance (outflow over inflow) ratio of the Nilo Coelho main canal (concrete lined, design capacity 20 m³/s, length 32.5 km)

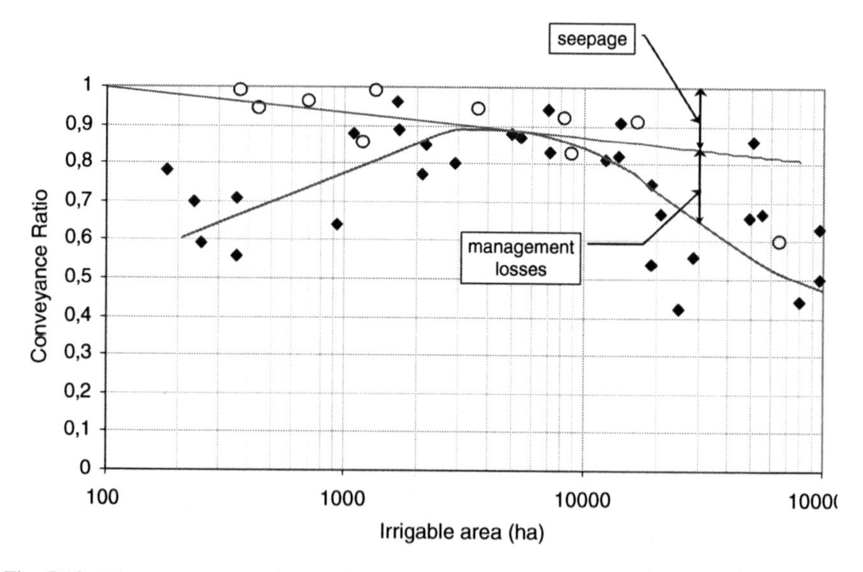

Fig. 5.12 The conveyance ratio as a function of the method of canal operation (continuous or intermittent flow) and the irrigable area (Bos and Nugteren 1974)

The curve shows a maximum R_c-value of about 0.9 for irrigable areas between 4,000 and 6,000 ha. For smaller areas, the R_c-value may be as low as 0.5, due to the reduction of project management to one person who, besides operating the irrigation system, is engaged in agricultural extension work, maintenance, transport and marketing of crops, administration, etc. If the manager is to perform all these tasks satisfactorily, he must be highly skilled. However, for small projects (i.e. less than 1,000 ha) funds are typically not available to hire such a person. Also if the irrigable area is large (more than 10,000 ha), the conveyance ratio decreases sharply due to problems that the management faces in controlling the water supply to small sub-areas. Large systems tend to be less flexible in adjusting the flow rate because of the relatively long time it takes to transmit information on actual flow rates and water requirements to a central office and the long travel time for water in open canals. However, in order to avoid water deficits in downstream canal sections, there is often a tendency to increase the flow (V_c) at the head of the canal system.

The operational performance of large systems can be improved considerably by reducing the effect of the above causes. The two most effective are:

- Divide the irrigated area into a number of lateral units, each having an area between 2,000 and 6,000 ha (depending on topography). Each lateral unit should receive its water (measured and controlled) at one point from the conveyance system and should have its own skilled local irrigation management staff that is responsible for the water distribution within that lateral unit only.
- Introduce a management performance programme that assesses the water delivery to all key points in the system. All these key points should be accurate flow control and measurement structure. The real-time data from these structures should be used for the operational (and strategic) water supply.

For water management within an existing irrigated area, we recommend that an initial target conveyance ratio be set by use of Fig. 5.12, and that the actual ratio be calculated on a monthly and an annual basis. As soon as dependable information is available on the monthly conveyance ratio, the total irrigation water requirements during the considered period (month) can be calculated.

5.5 Calculating Irrigation Water Requirements

Calculating the irrigation water requirements of a command area is a process that tracks two sorts of 'water flows' that should be managed in such a way that crops can be grown in the irrigable area (see Fig. 5.1):

- The classical flow of irrigation water from the surface water source (river diversion or reservoir) through the conveyance and distribution system to the fields. This track considers the 'crop water requirements' ($ET_p - P_e$) which subsequently is transferred into 'irrigation water requirements' for the irrigated crops (*ET*-track).
- The less visible (and often ignored) vertical flows of water seeping from the canals and fields to the groundwater basin and the 'return' flow through pumping and capillary rise. This track considers the water balance of the gross command area. It uses the depleted fraction (*DF*) as discussed in Chapter 4 in order to attain a sustainable irrigation environment (the *DF*-track).

Table 5.3 Parallel procedures used to estimate the irrigation water requirements of a command area

Water Requirements track for the irrigated crops	Water Balance track for the gross command area
For the current cropping pattern and the local meteorological conditions, use Chapter 2 in order to estimate the monthly potential evapotranspiration (ET_p) of all irrigated crops.	Establish the relationship between the depleted fraction and the monthly fluctuation of the groundwater table as discussed in Chapter 4. If no relationship is available assume $DF_{sust} = 0.67$.
Use any available literature in order to determine the crop production function for the major irrigated crop(s). Based on this function, decide on the range of the ratio ET_a/ET_p that is allowable for crop production. If no information is available, use the default range $0.7 \leq ET_a/ET_p \leq 1.0$.	
Based on the monthly precipitation values, select a strategy on related ET_a/ET_p values. Select lower value during wet months or during the mid- and late season of the major crop (consider the influence of the DF on the value of ET_a/ET_p).	Establish the relationship between the depleted fraction for the gross area and the ratio ET_a/ET_p for the irrigated area. If not available use $ET_a/ET_p = 1.0$ if $DF \leq DF_{sust}$, while ET_a/ET_p decreases to 0.6 if the DF increase to 1.0.
Calculate ET_a by multiplying the above ET_p with the ET_a/ET_p ratio that matches the selected DF-value.	Based on the monthly precipitation values, select a strategy on related DF-values. Aim at such a distribution that the annual average equals about DF_{sust}.
Based on the development of the effective rooting depth and the depth to the groundwater table, calculate the distance roots to groundwater. Using Chapter 4 on soil physical characteristics, calculate the rate of capillary rise into root zone.	Evaluate the anticipated monthly fluctuation of the groundwater table as function of the above selected strategy on the depleted fraction.
Calculate the 'crop irrigation water requirements' by subtracting the capillary rise and the effective precipitation (see Chapter 3) from the ET_a.	If no remote sensing data are available on $ET_{a,gross}$, use Chapter 2 in order to estimate the sum of ET_a (irrigated area) $ET_{a,fallow}$ (fallow irrigable area) and $ET_{a,non\text{-}ir}$ (permanently non-irrigated area).
Based on (the quality of) the irrigation water application method, select a 'field application ratio' and calculate $\sum V_f$ (total flow to all fields).	
Based on the quality of water delivery (uniformity) and the part of the area where 'water shortage' is acceptable, calculate $\sum V_f$ (total flow to all water delivery systems).	

(continued)

Table 5.3 (continued)

Water Requirements track for the irrigated crops	Water Balance track for the gross command area
Based on the size and the flow regime in the conveyance system, and on the quality of management decide on a 'conveyance ratio'. Then, calculate $V_{c,ET}$.	Using the definition of the depleted fraction: $DF = ET_{a,gross}/(V_{c,DF} + P)$, calculate the gross water requirement to maintain a sustainable water balance within the command area, $V_{c,DF}$.

↓

Review the above $V_{c,DF}$ and $V_{c,ET}$ values and the corresponding depleted fractions with respect to each other. In doing so, the following options are recommended:

↓

Value of $V_{c,DF}$ with respect to $V_{c,ET}$ and DF with respect to DF_{sust}	Recommendation
$V_{c,DF} \leq 0.95V_{c,ET}$ $DF \leq DF_{sust}$	Reduce $V_{c,ET}$ by improving the water delivery and/or the field water application.
$V_{c,DF} \leq 0.95V_{c,ET}$ $DF \geq DF_{sust}$	Allow a lower ET_a from the irrigated area and/or decrease the DF-value so that the value of $V_{c,DF}$ increases.
$0.95V_{c,ET} \leq V_{c,DF} \leq 1.05V_{c,ET}$ All values of DF	No recommendation; both water requirements are sufficiently close to each other.
$V_{c,DF} \geq 1.05V_{c,ET}$ $DF \leq DF_{sust}$	Apply water more uniformly so that ET_a from the irrigated area increases. Otherwise, increase the DF-value so that $V_{c,DF}$ decreases.
$V_{c,DF} \geq 1.05V_{c,ET}$ $DF \geq DF_{sust}$	Increase the DF-value for this month so that the value of $V_{c,DF}$ decreases.

Both tracks meet each other in an adjusted water management strategy as discussed at the bottom of Table 5.3. During the above two processes (tracks) we distinguish between the *consumption* and the *use* of water. If water is consumed (by the crop) it leaves the considered part of the system, and cannot be consumed or reused in other parts of the considered system. For example, if the field application ratio (efficiency) for a considered field is 65%; this means that 65% of the applied water is evapo-transpired and that the other 35% either becomes either surface run-off or recharges the aquifer.

During the irrigation process water can be used for a variety of other purposes than evapotranspiration by the crop. These uses may be directly related with irrigation (facilitate management, silt flushing, leaching, seepage, etc. see Fig. 5.13), or be related with other user groups (energy production, shipping, urban and industrial use, recreation, etc.). As a general rule, we may assume that the quality of water decreases upon its use. This section assumes that the water quality remains sufficient for re-use within the gross command area.

Fig. 5.13 View of the unlined section of the Cochella Canal showing vegetative growth in seepage area, August 1973 (Courtesy U.S. Bureau of Reclamation, E.E. Herzog)

Following adjustment and confirmation of the monthly value of $V_{c,ET}$ and $V_{c,DF}$, a water balance can be calculated (monthly and annually) using the highest of these V_c-values.

If the resulting annual average of the DF-value is lower than DF_{sust} the groundwater table will show an annual rise and artificial drainage is needed in order to control the groundwater table. The capacity of the artificial drainage system, V_{drain}, equals the total drainage from the command area minus the natural drainage capacity. Thus, on an annual basis it equals:

$$V_{drain} = \left(DF_{sust} - DF_{average}\right) \times \left(V_c + P\right). \quad 5.11$$

Chapter 6
Using the CRIWAR Software

6.1 Introduction

The previous chapters showed the importance of evapotranspiration as a component in the water balance of an irrigated area: Chapter 2 presents the theory and calculation procedures to estimate the potential evapotranspiration of agricultural crops. Chapter 3 presents three methods to estimate the contribution of precipitation to crop water requirements while Chapter 4 presents the water contribution to the effective root zone from capillary rise. Chapter 5 discusses the influence of irrigation technology and management strategies on water use. This last chapter combines all these water balance components into one simulation program. It describes the use of the CRIWAR software for developing water requirement tables and other output based on the selected water management strategy. CRIWAR 3.0 is the latest in a series of 'water requirement' analysis and evaluation programs developed since 1988. Table 6.1 summarizes the evolution of these computer programs.

6.2 Computer System Requirements

The CRIWAR 3.0 program has been compiled for 32-bit platforms for computers running Windows 98 and more recent operating systems.

6.3 Obtaining the Software

The current version of the CRIWAR software is maintained on the internet site www.bos-water.nl.

The program can be downloaded from this site, free of charge. CRIWAR 3.0 is public-domain and can be copied and freely distributed, as long as the software is not modified and appropriate recognition is given to its developers.

M.G. Bos et al. *Water Requirements for Irrigation and the Environment*,

Table 6.1 Computer software for irrigation water requirements (CRIWAR)

Version	Reference	Source code	Characteristics
1.0	Bos, M.G. 1988, Crop irrigation water requirements, ILRI, Wageningen (limited distribution)	Fortran	Estimated the potential evapotranspiration using the FAO modified Penman method.
2.0	Bos M.G., J. Vos, and R.A. Feddes. 1996. CRIWAR 2.0: A Simulation Model on Crop Irrigation Water Requirements. ILRI, Wageningen.	Turbo Pascal	Similar to CRIWAR 1.0 except that the Penman-Monteith method and a method to correct for effective precipitation were added. Theory and user manual were published.
3.0	Bos M.G., R.A.L. Kselik, R.J. Allen and D.J. Molden, 2008. Water Requirements for Irrigation and the Environment, Springer, Dordrecht.	Delphi 7.0	Interactive user interface on the irrigation water management strategy was added. The manual was upgraded to the level of M.Sc. level lecture notes for irrigation water management.

6.3.1 Installation

To begin the installation, you must extract the compressed file that you downloaded. To do so, execute the file you downloaded by double-clicking on it. Detailed installation instructions will be available from the CRIWAR web page on the Internet (see Section 6.3).

6.3.2 Starting the Program

Once installation is completed, CRIWAR may be started from the Windows Start Menu, or by double-clicking on the CRIWAR icon located in the CRIWAR program group.

6.4 Software Overview

CRIWAR maintains sub-directories as shown in Fig. 6.1. These sub-directories can be accessed through file handling programs. After starting the program you may load an existing file from the File menu, or create a new file definition using the *File* and *New File* commands.

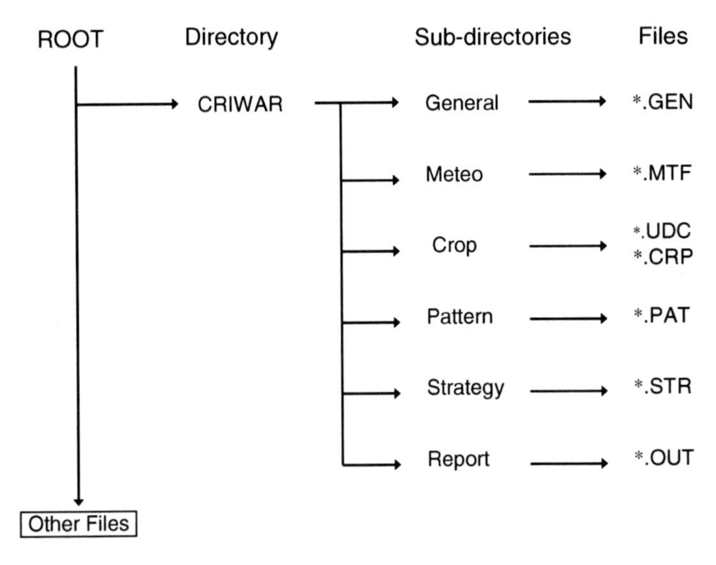

Fig. 6.1 Directory structure of CRIWAR

CRIWAR loads and saves files in its own file format with extensions as shown in Fig. 6.1. Files can be shared among users simply by copying the appropriate file to another user's computer.

To use CRIWAR, the user must open and complete at least three files containing *general* data, *meteo*rological data and a *cropping pattern* data for the irrigated command area (Fig. 6.2). Crops can be added to the cropping pattern if their characteristics are entered in a *crop factor* file. If these three files are opened, the reports and graphs on crop water requirement can be accessed. If only the *general data* and *meteorological* data are opened, the report will show only the reference evapotranspiration tab.

Under the *water balance strategy*, the user can develop a balance between the water demand of the irrigated crops and the environmental water demand for a stable groundwater table within the gross command area. Once this strategy has been developed, the related *reports and graphs* are accessible.

For the selection of a file, the user has three options. Either a new (blank) file can be opened and completed or an existing file can be opened. This existing file can either be used directly or can be modified.

6.5 Input Data Requirements

As mentioned above, CRIWAR requires files containing *general* data, *meteo*rological data and a *cropping pattern* data for the irrigated command area before the crop water requirement can be calculated. Once a file is selected, the related information is shown on the screen (see Fig. 6.3).

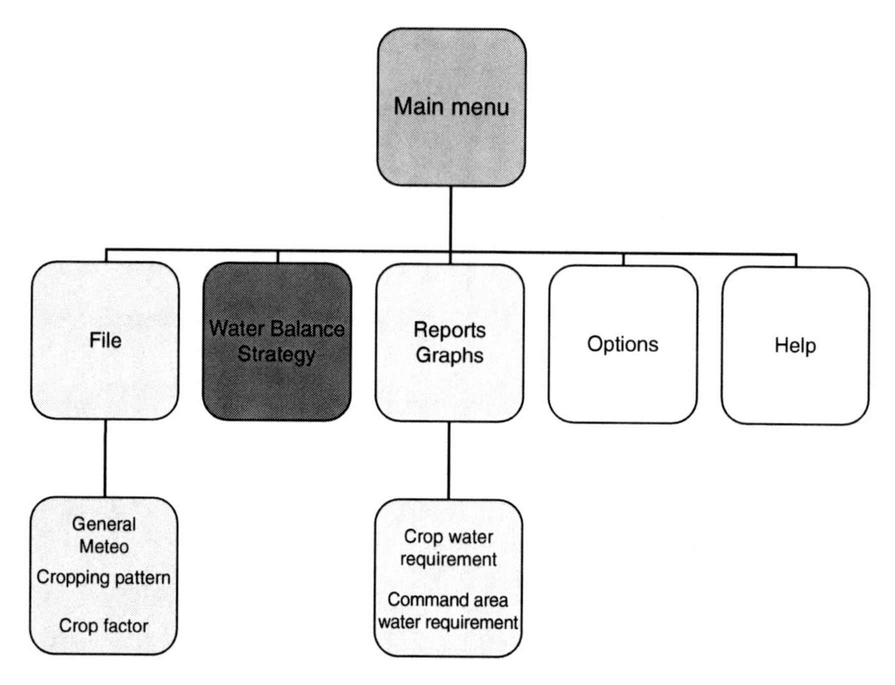

Fig. 6.2 Main menu structure

Criwar

File Water Balance Strategy Reports / Graphs Options Help

Data status

General	tunayan2.gen	Country	argentina	Description of irrigated area	Tunayan Middle Irrigation Syst
Meteo	tunayan2.mtf	Region / Project	Mendoza province	Size of irrigable area (ha)	4987.0
Cropping pattern	Tunayan.pat	Latitude	33.05 °South	Interval between applications (days)	20
Calculation period	Month	Altitude (m)	653.00	Mean depth of water application (mm)	75.0

Fig. 6.3 Information window on selected data files for the calculation of crop irrigation water requirements

6.5.1 *Entering General Data*

The general data screen is shown in Fig. 6.4. Data can be entered manually or by selecting from a list box. If the cursor is placed in a field, a range of acceptable values may be shown at the bottom of the screen. If values are entered outside this range, the value will be shown in 'red'. If no action is taken after this warning, CRIWAR will not generate any reports and graphs!

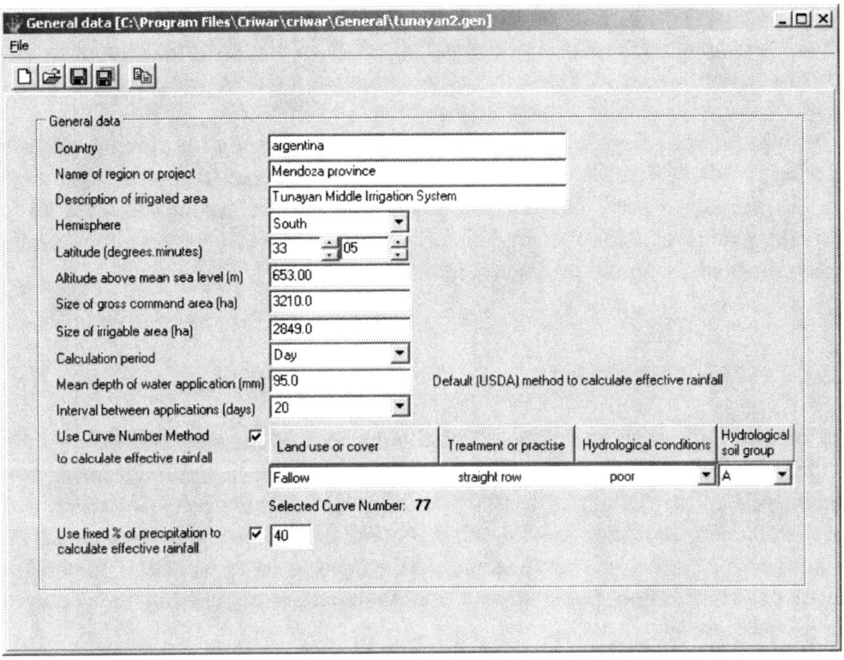

Fig. 6.4 The general data screen

6.5.1.1 Country, Name of Project, and Description of Area

These three options permit you to enter, or change, the location and the description of the irrigated area in the CRIWAR General Data File. These descriptions will be printed in the heading of the related tables. Hence, if you retrieve an existing file from the directory and subsequently edit it to meet new conditions, you should enter a new description.

6.5.1.2 Hemisphere, Latitude, and Altitude

In order to calculate the maximum possible number of hours of bright sunshine and the extra-terrestrial radiation, CRIWAR needs information on the hemisphere and the latitude of the irrigated area, and on its altitude above mean sea level. The hemisphere is defaulted 'North'.

6.5.1.3 Gross Command Area and Irrigable Area

The gross command area is the area within the outer boundary of the irrigation scheme. It includes non-irrigated areas like canals, roads and villages. The irrigable area is the area with physical infrastructure that enables the delivery of irrigation water. Often, the irrigable area is between 5% and 10% smaller than the gross command area. For sustainable water use within the gross command area, the water balance of this gross area is evaluated through the depleted fraction; $DF = ET_{a,gross}/(V_c + P)$.

The user-given irrigable area represents the 100% value for the cropping pattern. In other words, CRIWAR does not accept a larger cropped area under the menu branch *cropping pattern* than this user-given value. If the sum of sub-areas of the cropping pattern exceeds the irrigable area, CRIWAR will give a warning on the screen. Both areas should be entered in hectares (default) or in acres.

6.5.1.4 Calculation Period

The choice of the calculation period usually depends on the degree of detail of the available meteorological data and data on the cropping pattern. The data input period is defaulted 'Month'. The number of days in a calculation period varies with the month under consideration (28, 30, or 31 days). Other possibilities are to set the data input to a daily or a weekly value. Independent of the input setting, the output can be shown in different time-steps. The default output setting being the one set in the general screen.

6.5.1.5 Depth and Interval of Water Applications

The mean depth of irrigation water application per turn is used to estimate the effective part of the precipitation with the default USDA Method (see Section 3.3). For most irrigated areas, this application depth per turn would not exceed the readily available soil water in the root-zone. CRIWAR uses a default value of 75 mm. CRIWAR uses the frequency of an irrigation water application, or significant precipitation, in order to determine the crop coefficient, $K_{c,I}$ in the initial stage of crop growth (see Section 2.6.4).

6.5.1.6 Effective Rainfall

In order to calculate the effective precipitation, CRIWAR uses two semi-empirical methods. In addition, CRIWAR allows the user to set the effective precipitation as a fixed percentage of total precipitation. To generate output with the CN Method or the 'percentage method', the user must tick the related box in the *general data window*. The three methods are (see Chapter 3):

- The (default) method as developed by the U.S. Department of Agriculture (1970). This method can be used if monthly precipitation data is available. Output can only be generated in monthly time-steps. This method is described in Section 3.3.
- A method based on the Curve Number Method as developed by the U.S. Soil Conservation Service (1964 and 1972). This method requires daily precipitation

data. Hence, this method can be used only if the calculation period is set at 1 day. This method is described in Section 3.4.

- The user sets P_e as a percentage of P, while P_e cannot exceed ET_p during the considered calculation period.

6.5.2 Entering Meteorological Data

As already stated, CRIWAR can only calculate the crop irrigation water requirements for a region if meteorological data are available. Depending on the method selected to determine the reference evapotranspiration, data are needed as shown in Table 6.2. Depending on the selected method, the columns that need to be completed will be highlighted (Fig. 6.5). The meteo data table needs to be completed for the entire year.

As shown, data needs to be supplied on the day-night ratio of the wind speed and on the maximum relative humidity. However, if these data are not available, CRIWAR will use the following default values:

$$u_{day} / u_{night} = 2.0$$

and

$$RH_{max} = (RH + 100)/2$$

Where, RH is the average relative humidity. Day-time wind speed, u_{day}, is calculated from data on mean wind speed and a day-night wind ratio (see Section 2.2.2.2). Most numerical data have an allowable range of values (see Section 6.9 and the bottom of Fig. 6.5). If you input a value that is out of range, CRIWAR will show the related value(s) in **red**. A warning will be shown if you try to go to 'Reports and Graphs'.

Table 6.2 Data requirements for methods to estimate the reference evapotranspiration (Droogers and Allen 2002)

Data needed	FAO modified Penman	Penman-Monteith	Hargreaves-Samani
Minimum temperature	✓	✓	✓
Maximum temperature	✓	✓	✓
Humidity	✓	✓	
Wind speed	✓	✓	
Radiation	✓	✓	
Precipitation	☑	☑	✓

☑ meteorological data needed to calculate the effective precipitation

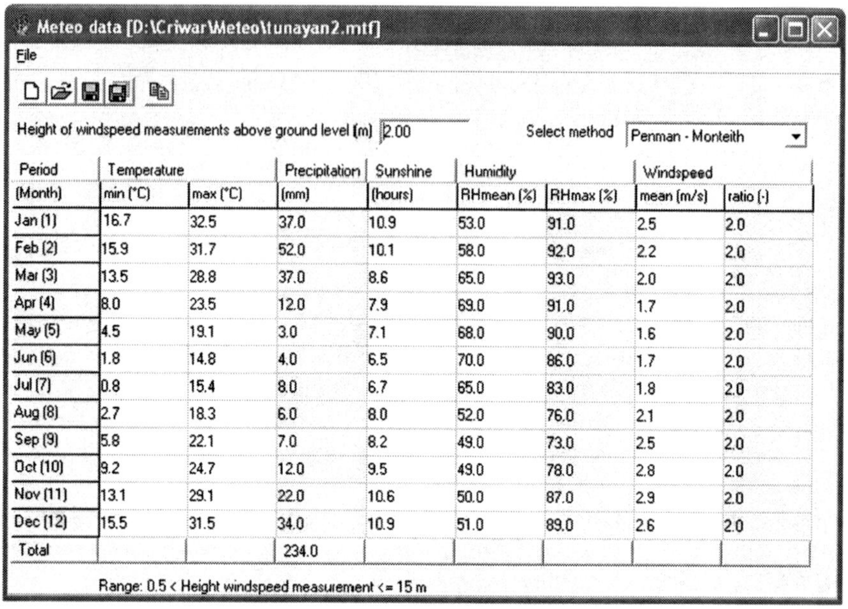

Meteo data [D:\Criwar\Meteo\tunayan2.mtf]

File

Height of windspeed measurements above ground level (m) 2.00 Select method Penman - Monteith

Period	Temperature		Precipitation	Sunshine	Humidity		Windspeed	
(Month)	min (°C)	max (°C)	(mm)	(hours)	RHmean (%)	RHmax (%)	mean (m/s)	ratio (-)
Jan (1)	16.7	32.5	37.0	10.9	53.0	91.0	2.5	2.0
Feb (2)	15.9	31.7	52.0	10.1	58.0	92.0	2.2	2.0
Mar (3)	13.5	28.8	37.0	8.6	65.0	93.0	2.0	2.0
Apr (4)	8.0	23.5	12.0	7.9	69.0	91.0	1.7	2.0
May (5)	4.5	19.1	3.0	7.1	68.0	90.0	1.6	2.0
Jun (6)	1.8	14.8	4.0	6.5	70.0	86.0	1.7	2.0
Jul (7)	0.8	15.4	8.0	6.7	65.0	83.0	1.8	2.0
Aug (8)	2.7	18.3	6.0	8.0	52.0	76.0	2.1	2.0
Sep (9)	5.8	22.1	7.0	8.2	49.0	73.0	2.5	2.0
Oct (10)	9.2	24.7	12.0	9.5	49.0	78.0	2.8	2.0
Nov (11)	13.1	29.1	22.0	10.6	50.0	87.0	2.9	2.0
Dec (12)	15.5	31.5	34.0	10.9	51.0	89.0	2.6	2.0
Total			234.0					

Range: 0.5 < Height windspeed measurement <= 15 m

Fig. 6.5 Example of a meteo window for monthly input data using the Penman-Monteith equation

6.5.2.1 Loading Meteo Data

The CRIWAR 3.0 program stores meteorological data in files, with an *.*MTF* extension (see Fig. 6.1). To avoid errors by re-typing, and to save time, tabulated data can be entered by 'copying and pasting' data from a spreadsheet into a meteo file.

6.5.3 Entering Cropping Pattern Data

Before CRIWAR can generate reports or graphs on water requirements, you first have to select a general data file, a meteo file, and a cropping pattern file to work with. If you attempt to select the menu branch **REPORTS/GRAPHS** before these files are selected, CRIWAR will require you to select those files first. Upon entering the **CROPPING PATTERN** branch of the menu, the window shown in Fig. 6.6 will appear on the screen:

Through the FILE menu, a database of example files is accessible. As before, there are three file options; open an existing file, edit an existing file or create a new file. In irrigated areas, studies often need to be made of the effect that changes in the cropping pattern will have on the irrigation water requirements. For this purpose, rather than create a new cropping pattern file for the area, you will simply want to modify its existing cropping pattern file. If, for some reason, no existing cropping pattern file (named *.PAT) is available, a new pattern file can be created. There are many situations where you may want to compose such an entirely new cropping pattern. This is particularly true for newly designed irrigation systems. Equally, there are many situations where a calibration is needed, based on the 'as-measured' cropping pattern of an existing irrigated area.

Figure 6.6 shows which crops are loaded into the cropping pattern. Data on *crop* name, *growing period* and *description of crop* are imported from the CROP FACTOR FILE and cannot be changed in this screen. The *cropped area* and *planting month/ date* are entered, or imported, through the ADD (OR EDIT) CROP TO PATTERN windows. CRIWAR keeps track of the intensity of land occupancy by crops in the irrigable area. A warning will be shown if the land occupancy is larger than the irrigable area.

To delete a crop from the pattern in Fig. 6.6, highlight the crop by putting the cursor on the row with the crop and click the × button. To add a crop, click the + button upon which the window of Fig. 6.7 is shown. The blank window should be completed (or edited) in two steps:

Firstly, a crop needs to be selected from existing *crop factor files* by selecting a crop through the *File Open* button. Hereby, two sorts of files can be loaded: a CRIWAR pre-defined file (*.crp) and a user defined file (*.udc). Only existing files can be loaded. The method by which user defined crop factor files can be made is discussed in Section 6.5.4.

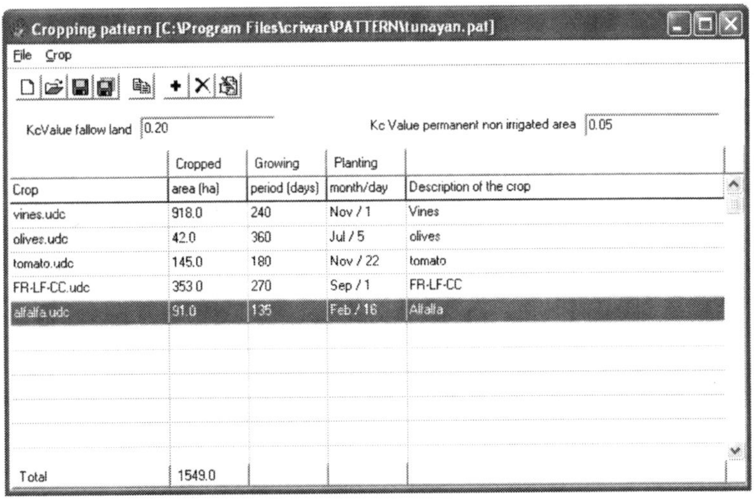

Cropping pattern [C:\Program Files\criwar\PATTERN\tunayan.pat]

File Crop

KcValue fallow land 0.20 Kc Value permanent non irrigated area 0.05

Crop	Cropped area (ha)	Growing period (days)	Planting month/day	Description of the crop
vines.udc	918.0	240	Nov / 1	Vines
olives.udc	42.0	360	Jul / 5	olives
tomato.udc	145.0	180	Nov / 22	tomato
FR-LF-CC.udc	353.0	270	Sep / 1	FR-LF-CC
alfalfa.udc	91.0	135	Feb / 16	Alfalfa
Total	1549.0			

Fig. 6.6 Example of a cropping pattern file

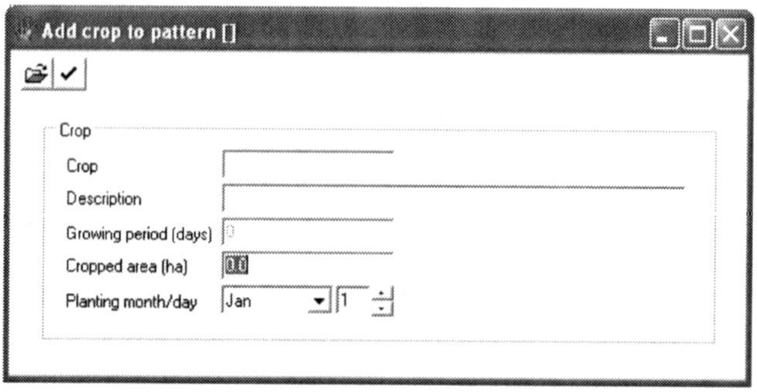

Add crop to pattern []

Crop

Crop

Description

Growing period (days) 0

Cropped area (ha) 0.0

Planting month/day Jan 1

Fig. 6.7 Window used to add crop or edit crop area and planting date

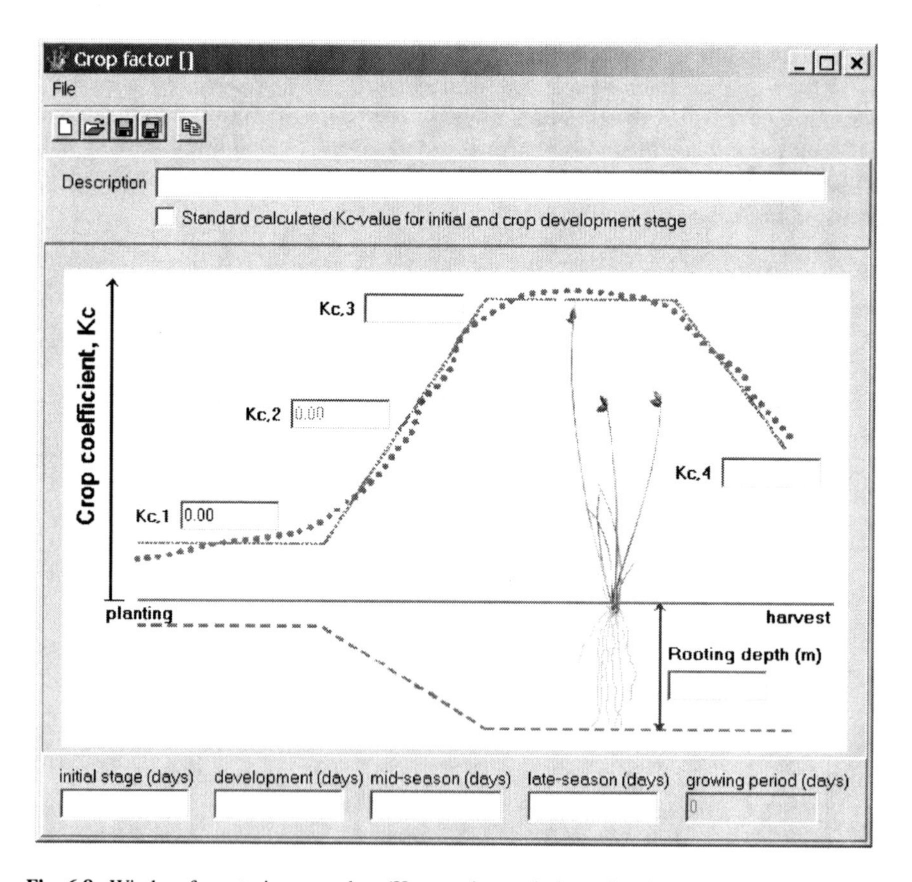

Fig. 6.8 Window for entering crop data (K_c, growing period, rooting depth)

Secondly, the cropped area and the planting month/date need to be entered. As mentioned, these data are used to keep track of the intensity of land occupancy by crops in the irrigable area. If the occupancy exceeds 100% on any day, a warning will be shown as soon as the ☑ button is clicked.

6.5.3.1 K_c-Value for Fallow Land and Non-irrigated Area

The cropping pattern window accepts two K_c-values for fallow land and permanently non-irrigated area respectively. These values are not used to calculate the irrigation water requirement, but to estimate the depleted fraction of the gross command area (see Sections 2.9 and 6.7)

6.5.4 Entering Crop File Data

As mentioned in Section 6.5.3, two sorts of existing crop factor files can be loaded into a cropping pattern: CRIWAR pre-defined crops (*.crp) or user defined crops with the extension (*.udc). A selection of CRIWAR pre-defined files is available. Files can be deleted only via Windows Explorer (see Fig. 6.1 for file locations). Files can be edited under the CROP FILES menu. The related window shows a schematic growth curve of a crop and its effective rooting system (Fig. 6.8).

The value of the crop coefficient during the crop's initial and development stages can either be calculated by CRIWAR or be user defined (default). If you want CRIWAR to calculate these values tick the related checkbox for the standard calculated initial Kc value. The two related k_c fields become inaccessible, and will show the CRIWAR-calculated values (see Section 2.7.1). CRIWAR will calculate the total length of the growing period, and will write this length in the cropping pattern table. After you have entered data in all fields, you can save the file for future use. Information on the crop will be shown in the last column of the cropping pattern table.

In order to estimate the capillary contribution of water, we need information about the effective rooting depth of the crops in the area. For this we consider the following rooting development (Section 4.5):

- During the initial growth stage the crop just has been planted or seeded. For all crops we assume an effective rooting depth of 0.05 m.
- During crop development the above-ground biomass production grows in proportion with the root system. We assume a linear rooting development during this stage.
- During the mid-season and late-season growth stages the effective rooting depth of the crop is assumed to be constant. It can be estimated from Table 4.2.

Please note, that an edited crop file becomes a '**u**ser-**d**efined **c**rop' factors file (named *.UDC) and that you cannot overwrite a CRIWAR pre-programmed file. You can only delete these protected crop files by using Window™ commands. To edit a user-defined crop (factors file) and add it to the cropping pattern, you need to follow the aforementioned procedure. At present, however, you are able to overwrite an existing user-defined crop factors file.

If a crop factor file is edited while the crop is already selected in a cropping pattern, this pattern should be reloaded before the changes will take place.

6.6 Developing a Water Management Strategy

CRIWAR views the water requirements of an irrigated area from two perspectives; (1) the water requirements for a sustainable water balance in the gross command area and (2) the water requirements of the irrigated crop. Both strategies result to an irrigation water demand. These two demands are subsequently matched.

6.6.1 Strategy for a Sustainable Water Balance

The water balance strategy is based on the relationship between the depleted fraction of the gross command area, $ET_{a,gross}/(V_c + P)$, and the fluctuation of the groundwater table (see Section 4.4) and on the relationship between this depleted fraction and the relative evapotranspiration, ET_a/ET_p, in the irrigated part of the command area. There are two options for you to deal with the water balance:

- No measured data are available on groundwater fluctuation and on the inputs needed to calculate the depleted fraction. In this case default relationships will be used as shown in Fig. 6.9. The only value that can be changed in this window is the 'sustainable value' of the depleted fraction, DF_{sust} (see Section 4.6 for values). The 'groundwater line' is used to estimate monthly fluctuation of the groundwater table as a function of the user-selected strategy on the depleted fraction. The '*ET*-curve' is used to check on consistency between the water balance strategy and the *ET*-strategy.
- If the field data available box is checked, the window of Fig. 6.9 turns blank and a data input tab becomes available. In this window (Fig. 6.10) measured field data can be entered: monthly change of groundwater table, diverted irrigation water, V_c, and the actual evapotranspiration from the gross command area, $ET_{a,gross}$. The shown precipitation comes from the selected meteo file while the depleted fraction is calculated using the entered data. As with other windows, existing files can be loaded and edited if needed. As soon as data are

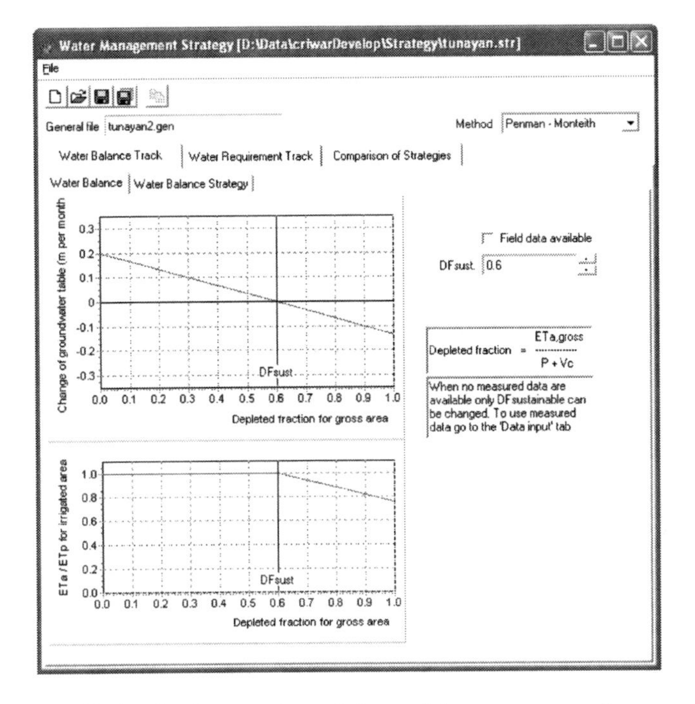

Fig. 6.9 Default Window for the groundwater balance of the gross command area

entered in the window of Fig. 6.10, the 'groundwater line' of Fig. 6.9 will be fitted through the data points. The calculated value of DF_{sust} will be used to shift the *ET*-curve to the related position. Note, that the shape of this curve will not change. Following the entering of all data, the *water balance* window looks like the example of Fig. 6.11. The entered data can be saved in a file named *.str.

Under the *water balance strategy* tab (Fig. 6.12), the user can enter or edit the depleted fraction in such a way that the annual change of the groundwater table remains near zero. The monthly precipitation (from the meteo file) is shown in order to assist with this strategy. The calculated value of $V_{c,DF}$ is the water requirement in order to maintain a stable groundwater table within the gross command area. With this volume of water there is neither water logging nor groundwater mining. Hence, $V_{c,DF}$ is the volume of water needed for a sustainable environment. It is calculated from the user-selected depleted fraction (*DF*) as:

$$V_{c,DF} = \frac{ET_{a,gross}}{DF} - P \qquad 6.1$$

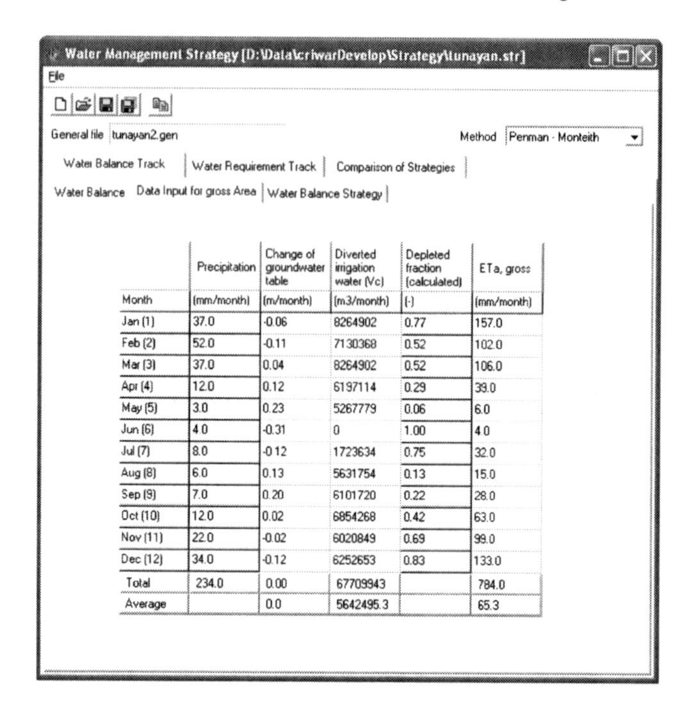

Water Management Strategy [D:\Data\criwarDevelop\Strategy\tunayan.str]

File

General file tunayan2.gen Method Penman - Monteith

Water Balance Track | Water Requirement Track | Comparison of Strategies

Water Balance | Data Input for gross Area | Water Balance Strategy

Month	Precipitation (mm/month)	Change of groundwater table (m/month)	Diverted irrigation water (Vc) (m3/month)	Depleted fraction (calculated) (-)	ETa, gross (mm/month)
Jan (1)	37.0	-0.06	8264902	0.77	157.0
Feb (2)	52.0	-0.11	7130368	0.52	102.0
Mar (3)	37.0	0.04	8264902	0.52	106.0
Apr (4)	12.0	0.12	6197114	0.29	39.0
May (5)	3.0	0.23	5267779	0.06	6.0
Jun (6)	4.0	-0.31	0	1.00	4.0
Jul (7)	8.0	-0.12	1723634	0.75	32.0
Aug (8)	6.0	0.13	5631754	0.13	15.0
Sep (9)	7.0	0.20	6101720	0.22	28.0
Oct (10)	12.0	0.02	6854268	0.42	63.0
Nov (11)	22.0	-0.02	6020849	0.69	99.0
Dec (12)	34.0	-0.12	6252653	0.83	133.0
Total	234.0	0.00	67709943		784.0
Average		0.0	5642495.3		65.3

Fig. 6.10 Window for data input on water balance of the gross command area

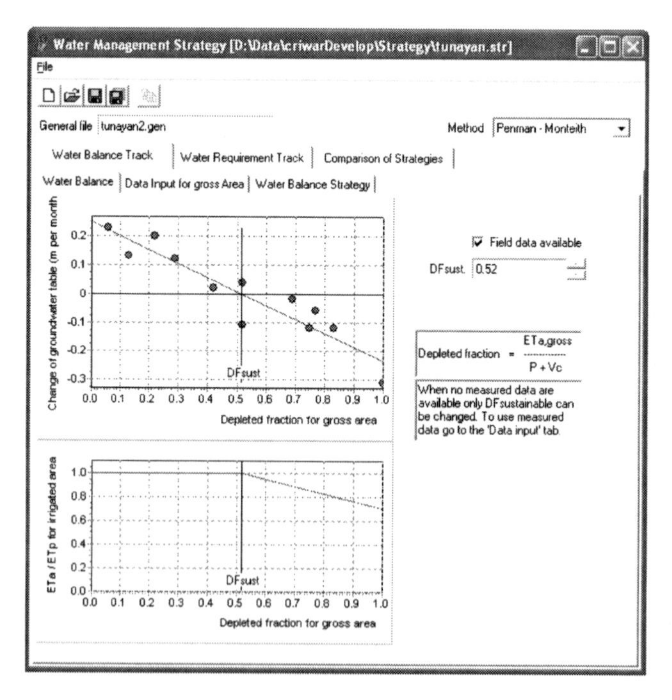

Fig. 6.11 Water balance window if filed data are available on groundwater fluctuation

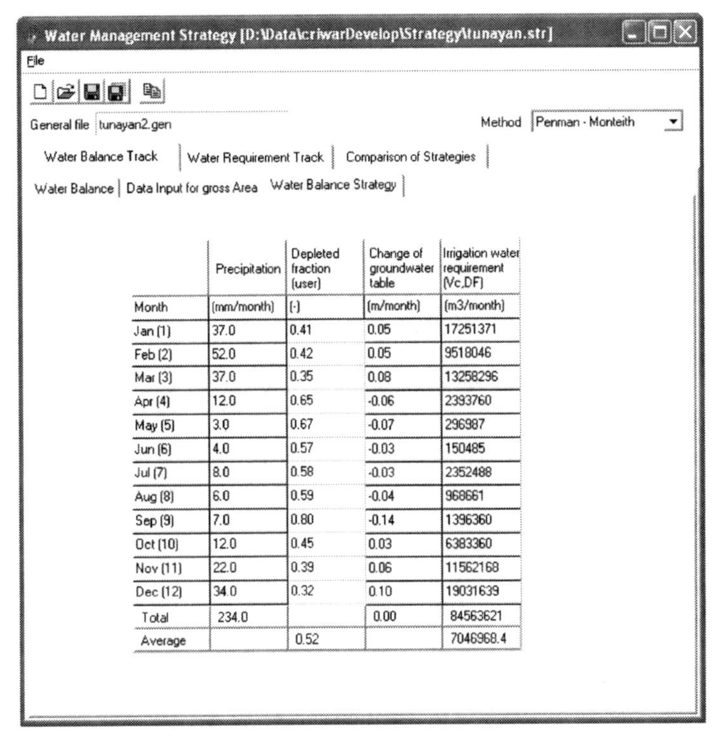

	Precipitation	Depleted fraction (user)	Change of groundwater table	Irrigation water requirement (Vc,DF)
Month	(mm/month)	(-)	(m/month)	(m3/month)
Jan (1)	37.0	0.41	0.05	17251371
Feb (2)	52.0	0.42	0.05	9518046
Mar (3)	37.0	0.35	0.08	13258296
Apr (4)	12.0	0.65	-0.06	2393760
May (5)	3.0	0.67	-0.07	296987
Jun (6)	4.0	0.57	-0.03	150485
Jul (7)	8.0	0.58	-0.03	2352488
Aug (8)	6.0	0.59	-0.04	968661
Sep (9)	7.0	0.80	-0.14	1396360
Oct (10)	12.0	0.45	0.03	6383360
Nov (11)	22.0	0.39	0.06	11562168
Dec (12)	34.0	0.32	0.10	19031639
Total	234.0		0.00	84563621
Average		0.52		7046968.4

Fig. 6.12 Window on the user-defined water balance strategy

The value of $ET_{a,gross}$ either equals the measured value as entered by the user in Fig. 6.10 or is estimated from the equation:

$$ET_{a,gross} = C_{a/p}\left(ET_p + ET_{p,non.ir} + ET_{p,fallow}\right) \qquad 6.2$$

Where,

$C_{a/p}$ = a correction coefficient converting potential into actual evapotranspiration. It equals 1.0 if the depleted fraction is less than DF_{sust}. For higher *DF*-values it decreases as shown in Fig. 6.9.

ET_p = the potential evapotranspiration for the irrigated area as calculated by CRIWAR

$ET_{p,non.ir}$ = the potential evapotranspiration from the permanently non-irrigated area within the gross command area (see Section 2.9)

$ET_{p,fallow}$ = the potential evapotranspiration from the fallow land within the irrigated area (see Section 2.9)

6.6.2 Strategy on Water Requirement of the Irrigated Crop

Calculating the irrigation water requirement from the crops' perspective starts with the CRIWAR calculated value of crop irrigation water requirement for non-stressed conditions, being equal to ($ET_p - P_e$). In this crop water requirement, the effective precipitation (P_e) is calculated as described in Chapter 3. As discussed in Chapter 5, the value of ET_a can be somewhat less than ET_p without significantly reducing crop yield. Selecting a water supply strategy whereby a lower ET_a is accepted during dry (water scarce) months can improve the crop productivity in terms of water; e.g. crop yield in terms of water consumed (thus in kg/m^3 water).

The window shown in Fig. 6.13 allows the user to enter this strategy. In this window, the precipitation comes from the user-selected meteo file, while the ET_p for the irrigated cropping pattern was calculated before. The ratio ET_a/ET_p is calculated immediately after data entry.

Besides precipitation, the field (crops) receives water through capillary rise and through irrigation. Capillary rise into the effective root zone depends on the *dominant soil type* and on the distance between the groundwater table and the 'bottom' of the effective root zone. The dominant soil type can be selected from a drop down

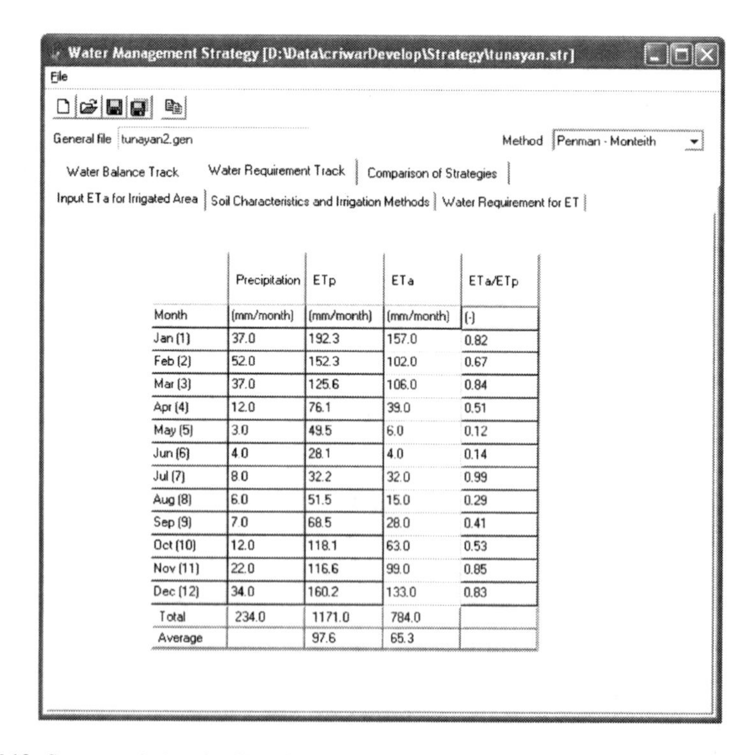

	Precipitation	ETp	ETa	ETa/ETp
Month	(mm/month)	(mm/month)	(mm/month)	(-)
Jan (1)	37.0	192.3	157.0	0.82
Feb (2)	52.0	152.3	102.0	0.67
Mar (3)	37.0	125.6	106.0	0.84
Apr (4)	12.0	76.1	39.0	0.51
May (5)	3.0	49.5	6.0	0.12
Jun (6)	4.0	28.1	4.0	0.14
Jul (7)	8.0	32.2	32.0	0.99
Aug (8)	6.0	51.5	15.0	0.29
Sep (9)	7.0	68.5	28.0	0.41
Oct (10)	12.0	118.1	63.0	0.53
Nov (11)	22.0	116.6	99.0	0.85
Dec (12)	34.0	160.2	133.0	0.83
Total	234.0	1171.0	784.0	
Average		97.6	65.3	

Fig. 6.13 Strategy window for ET_a of the irrigated area

list in Fig. 6.14 upon which the *height of capillary rise* is shown for three flow rates (fluxes). The three fluxes are based on the example soils from Fig. 4.4. The user can edit the fluxes if needed.

The lower part of the window deals with the method of irrigation at three levels: field, distribution system and conveyance system. Drop down lists give methods and related data. All data are based on the theory of Chapter 5. The user can edit data in order to match the local conditions.

The following tab (Fig. 6.15) allows you to enter the initial depth to the groundwater table on 1 January (i.e. 1st Julian day). CRIWAR then uses the depleted fraction to estimate the monthly fluctuation of the groundwater table. Since the development of the effective rooting depth for each crop is estimated (Section 4.5), CRIWAR can calculate the distance between the groundwater table and the effective root zone. Based on this distance and on the soil characteristics, the upward capillary flow is estimated. Linear interpolation is used in order to calculate values for each Julian day. The last column of Fig. 6.15 gives the project water requirement in accordance with the crop irrigation water requirement (see also Section 5.2).

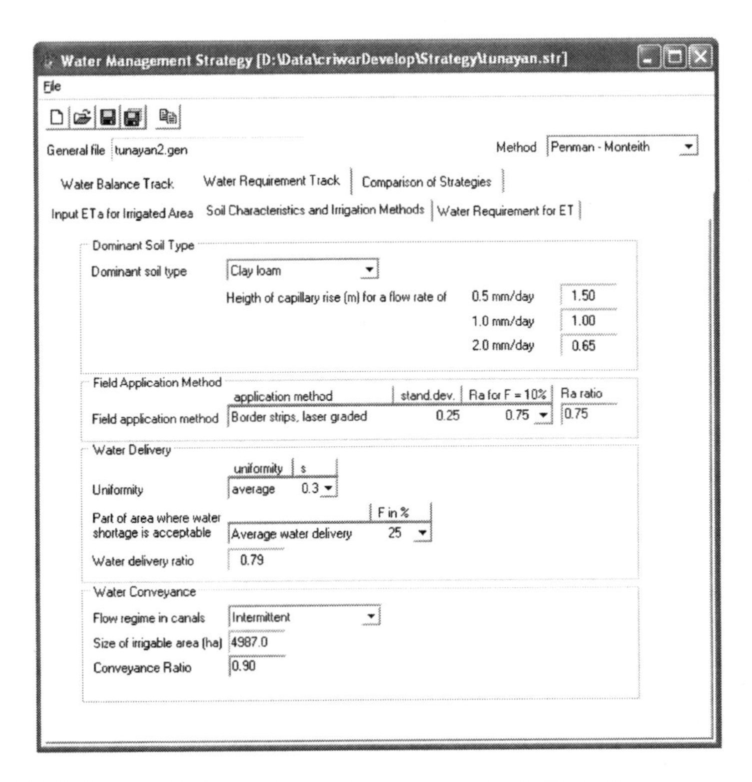

Fig. 6.14 Window on soil characteristics and the chosen method of irrigation

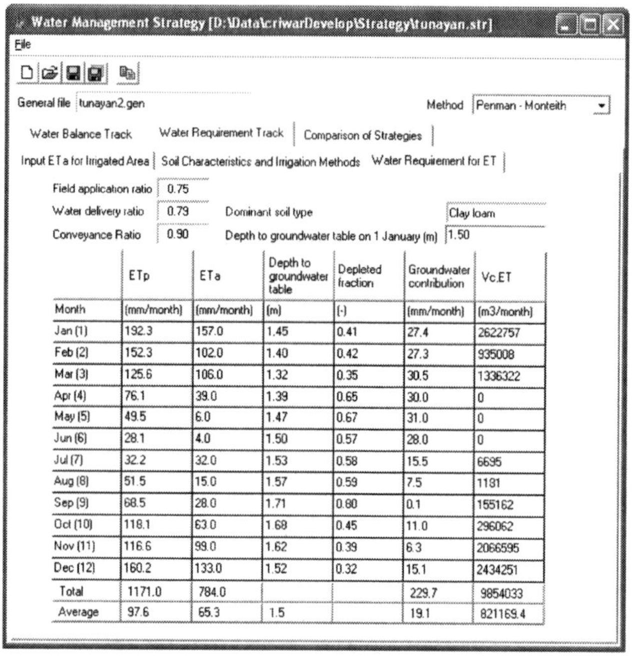

Water Management Strategy [D:\Data\criwarDevelop\Strategy\tunayan.str]

File

General file tunayan2.gen

Method Penman - Monteith

Water Balance Track | Water Requirement Track | Comparison of Strategies

Input ETa for Irrigated Area | Soil Characteristics and Irrigation Methods | Water Requirement for ET

Field application ratio 0.75

Water delivery ratio 0.79 Dominant soil type Clay loam

Conveyance Ratio 0.90 Depth to groundwater table on 1 January (m) 1.50

	ETp	ETa	Depth to groundwater table	Depleted fraction	Groundwater contribution	Vc.ET
Month	(mm/month)	(mm/month)	(m)	(-)	(mm/month)	(m3/month)
Jan (1)	192.3	157.0	1.45	0.41	27.4	2622757
Feb (2)	152.3	102.0	1.40	0.42	27.3	935008
Mar (3)	125.6	106.0	1.32	0.35	30.5	1336322
Apr (4)	76.1	39.0	1.39	0.65	30.0	0
May (5)	49.5	6.0	1.47	0.67	31.0	0
Jun (6)	28.1	4.0	1.50	0.57	28.0	0
Jul (7)	32.2	32.0	1.53	0.58	15.5	6695
Aug (8)	51.5	15.0	1.57	0.59	7.5	1181
Sep (9)	68.5	28.0	1.71	0.80	0.1	155162
Oct (10)	118.1	63.0	1.68	0.45	11.0	296062
Nov (11)	116.6	99.0	1.62	0.39	6.3	2066595
Dec (12)	160.2	133.0	1.52	0.32	15.1	2434251
Total	1171.0	784.0			229.7	9854033
Average	97.6	65.3	1.5		19.1	821169.4

Fig. 6.15 Review of groundwater contribution to the crop water requirement

6.6.3 Analyzing Alternative Water Management Strategies

Once the two alternative water strategies have been completed (i.e. after the user has entered initial values of the monthly depleted fraction and the monthly relative ET), the strategy evaluation window can be used to compare alternative water requirements (Fig. 6.16). Note that the two strategies will have different values of $V_{c,ET}$ and $V_{c,DF}$. Ideally, the two values should be the same order of magnitude: differ less than about 5% from each other. A better match can be made by changing the $V_{c,DF}$ and/or $V_{c,ET}$ values. To assist in this process, the relevant options of Table 6.3 are listed in the last column of the window.

The window of Fig. 6.16 merely serves to compare the current $V_{c,ET}$ and $V_{c,DF}$ values. Strategy changes cannot be made in this window. For this, you should return the related part of the software as described in Sections 6.6.1 and 6.6.2.

The *water management strategy* part of CRIWAR only addresses the water management questions and irrigation technology options as discussed in Chapter 5. Other refinements of the irrigation water requirement can be made in the *crop water requirement* part of CRIWAR (e.g., adjustments to the type and variety of irrigated crops, irrigated area per crop, selected K_c coefficient, etc.).

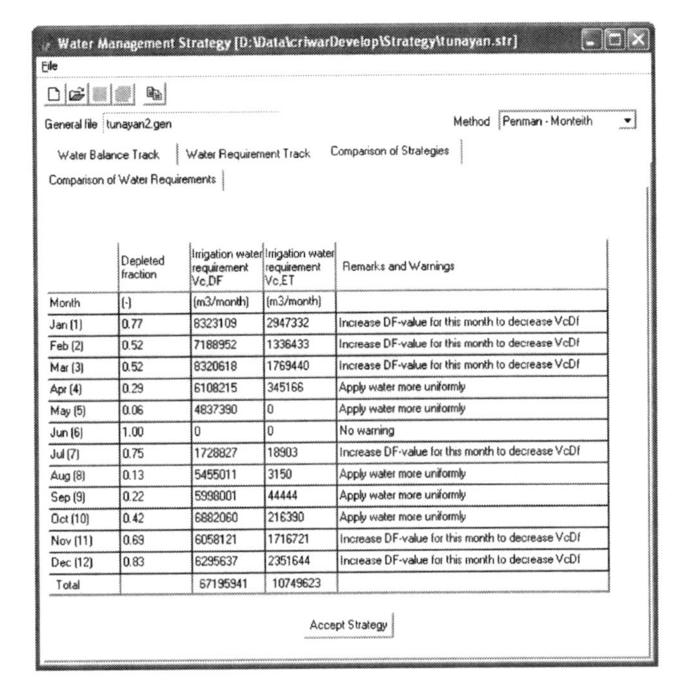

Month	Depleted fraction (-)	Irrigation water requirement Vc,DF (m3/month)	Irrigation water requirement Vc,ET (m3/month)	Remarks and Warnings
Jan (1)	0.77	8323109	2947332	Increase DF-value for this month to decrease VcDf
Feb (2)	0.52	7188952	1336433	Increase DF-value for this month to decrease VcDf
Mar (3)	0.52	8320618	1769440	Increase DF-value for this month to decrease VcDf
Apr (4)	0.29	6108215	345166	Apply water more uniformly
May (5)	0.06	4837390	0	Apply water more uniformly
Jun (6)	1.00	0	0	No warning
Jul (7)	0.75	1728827	18903	Increase DF-value for this month to decrease VcDf
Aug (8)	0.13	5455011	3150	Apply water more uniformly
Sep (9)	0.22	5998001	44444	Apply water more uniformly
Oct (10)	0.42	6882060	216390	Apply water more uniformly
Nov (11)	0.69	6058121	1716721	Increase DF-value for this month to decrease VcDf
Dec (12)	0.83	6295637	2351644	Increase DF-value for this month to decrease VcDf
Total		67195941	10749623	

Fig. 6.16 Comparison of two water demands: $V_{c,DF}$ and $V_{c,ET}$

Table 6.3 Recommended changes in water use strategy

Remark#	Value of $V_{c,DF}$ with respect to $V_{c,ET}$ and DF with respect to DF_{sust}	Recommendation
1	$V_{c,DF} \leq 0.95V_{c,ET}$ $DF \leq DF_{sust}$	Reduce $V_{c,ET}$ by improving the water delivery and/or the field water application.
2	$V_{c,DF} \leq 0.95V_{c,ET}$ $DF \geq DF_{sust}$	Allow a lower ET_a from the irrigated area and/or decrease the *DF*-value so that the value of $V_{c,DF}$ increases.
3	$0.95V_{c,ET} \leq V_{c,DF} \leq 1.05V_{c,ET}$ All values of *DF*	No recommendation; both water requirements are sufficiently close to each other.
4	$V_{c,DF} \geq 1.05V_{c,ET}$ $DF \leq DF_{sust}$	Apply water more uniformly so that ET_a from the irrigated area increases. Otherwise, increase the *DF*-value so that $V_{c,DF}$ decreases.
5	$V_{c,DF} \geq 1.05V_{c,ET}$ $DF \geq DF_{sust}$	Increase the *DF*-value for this month so that the value of $V_{c,DF}$ decreases.

Following adjustment of the strategies, and confirmation of the monthly value of $V_{c,ET}$ and $V_{c,DF}$, a water balance will be calculated (monthly and annual) using the highest of these two V_c-values. Where, outputs can be generated under *Reports and Graphs*.

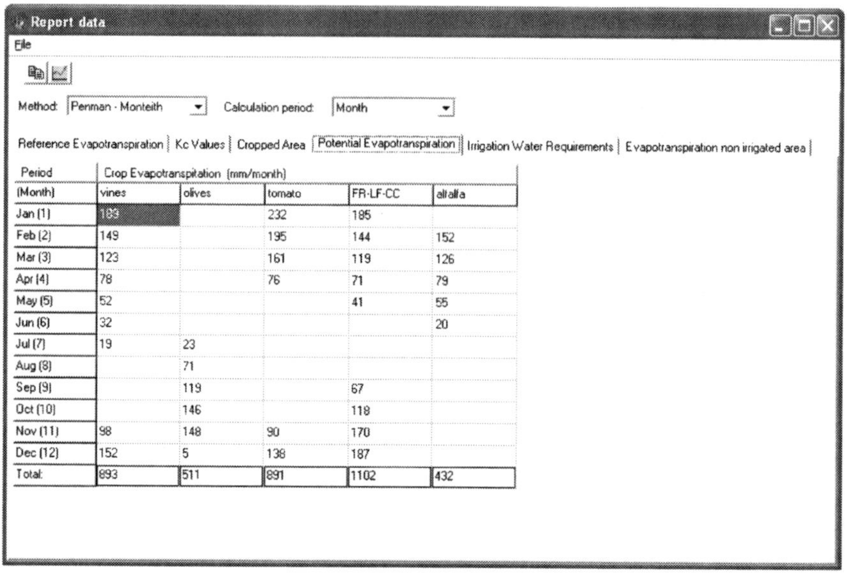

Report data

File

Method: Penman - Monteith Calculation period: Month

Reference Evapotranspiration | Kc Values | Cropped Area | Potential Evapotranspiration | Irrigation Water Requirements | Evapotranspiration non irrigated area

Period	Crop Evapotranspitation (mm/month)				
(Month)	vines	olives	tomato	FR-LF-CC	alfalfa
Jan (1)	189		232	185	
Feb (2)	149		195	144	152
Mar (3)	123		161	119	126
Apr (4)	78		76	71	79
May (5)	52			41	55
Jun (6)	32				20
Jul (7)	19	23			
Aug (8)		71			
Sep (9)		119		67	
Oct (10)		146		118	
Nov (11)	98	148	90	170	
Dec (12)	152	5	138	187	
Total:	893	511	891	1102	432

Fig. 6.17 Tabbed output window on crop water requirements

6.7 Producing Output

Several types of output are available from the program, including graphs and tables. All output from the program can be printed to any Windows-compatible printer. All program output can be copied to the system clipboard for pasting into Windows-compatible word processors or spreadsheets.

6.7.1 Tables & Graphs

Several types of tables are available in the program. All tables are obtained through the tabbed table form shown in Fig. 6.17. A water requirement table shows the discharges corresponding to fixed intervals of the water balance unit (e.g. m^3/s, m^3/day, m^3/week, m^3/month or m^3/year).

The *graph* button of the Tables & Graphs form shows the table data in graphical form (Fig. 6.18). When the chart is saved, the image is stored in the Folder reports. From there it can be loaded into a document or spreadsheet.

6.8 Program Options

Several miscellaneous program options can be set from the Options menu:

User name

Units

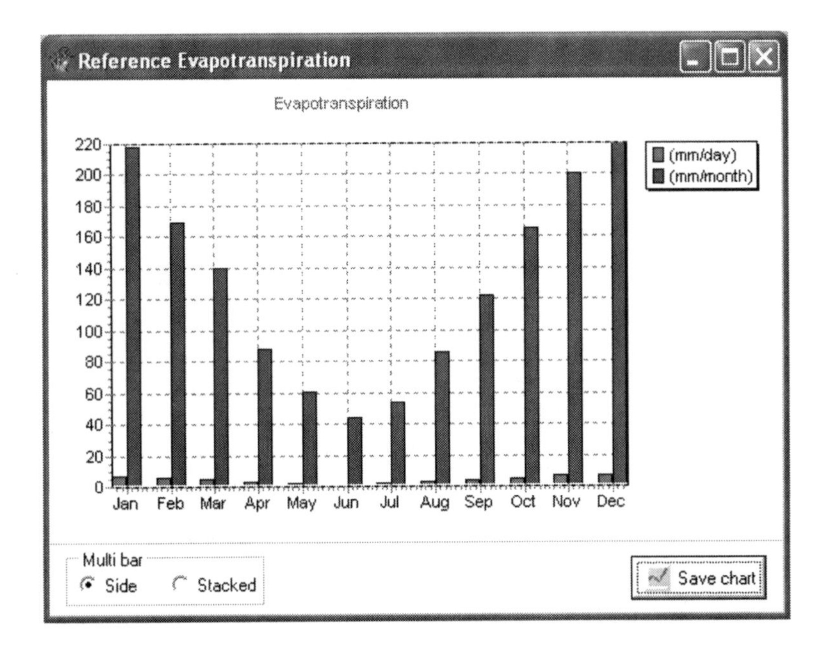

Fig. 6.18 Combinations of two parameters can be selected for presentation in a graph

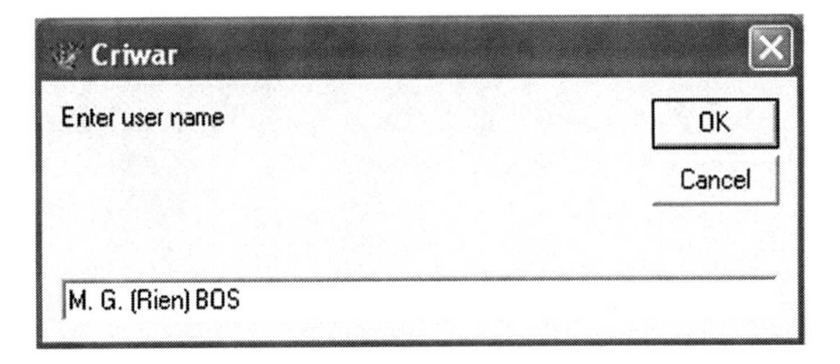

Fig. 6.19 Window for entering user name

6.8.1 User Name

Prompts the user for their name (Fig. 6.19). This user name becomes a setting of this CRIWAR installation and will be printed on all reports produced by the program. The user name will be saved and recalled for use in future CRIWAR sessions. The user name can be changed at any time by selecting User Name from the Options menu.

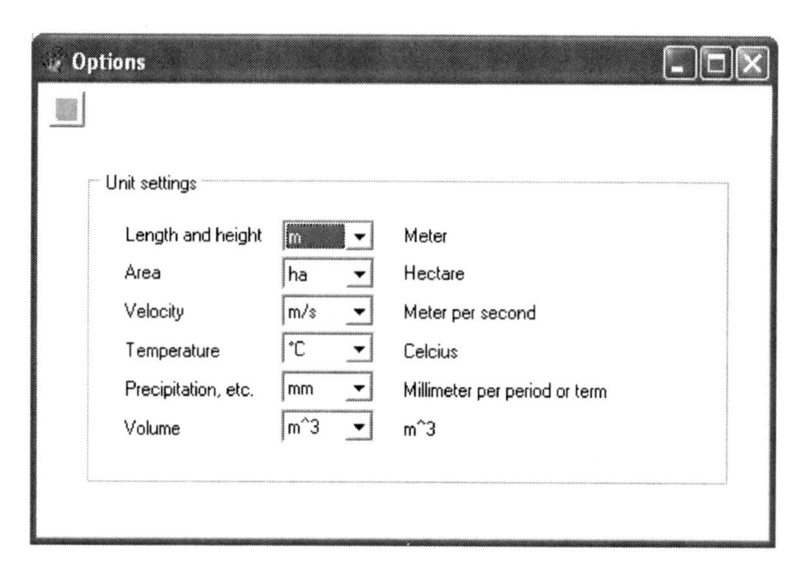

Fig. 6.20 Window for setting units

6.8.2 Choose Units System

CRIWAR works with and stores all files internally using SI units, but you can choose other units for display and input of General Data, Meteo Data and Water Management Strategy. The dialog box shown in Fig. 6.20 allows you to choose the system of units used for displaying and entering area, length, velocity, and temperature.

6.9 Warnings and Error Messages

CRIWAR only accepts input parameters that have a numerical value within a pre-programmed range. If you enter an out-of-range value, CRIWAR will show that value in red. The relevant range is given in the bottom of the input window. You should select a value that is within the valid range. Tables 6.4 and 6.5 show the range of values used by CRIWAR and the dimensions of the various numerical parameters.

6.9.1 Additional Warnings

6.9.1.1 Name of Crop File

The same crop name can be selected more than once. This has the advantage that more crops of the same type can be planted at different dates in one pattern.

Table 6.4 Range of values of meteorological input parameters

Parameter	Range			Units
Latitude	0	to	66	degrees north or south
Altitude	−500	to	4,500	m
Height of wind speed measured	0	to	15	m
Temperature	0	to	55	degrees °C
Precipitation	0	to	1,000	mm/period
Sunshine hours	0	to	24	h/day
Relative humidity	0	to	100	%
Wind speed	0	to	15	m/s
Maximum relative humidity	rhu_{min}	to	100	%
Wind speed ratio day/night	0	to	5	dimensionless

Table 6.5 Range of acceptable values for crop-related input parameters

Parameter			Range	Units
Total cropped area	1	to	1,000,000	ha
Mean application depth of water	20	to	200	mm/turn
Crops in cropping pattern	0	to	40	dimensionless
Crop coefficient of user given crop	0	to	2.0	dimensionless

6.9.1.2 Total Irrigated Area

While the cropping pattern is being composed, CRIWAR will keep track of the total irrigated area per month (or user selected period). If this total exceeds the user-given total irrigable area of the selected general data file, CRIWAR will display a warning. Since no irrigable area remains available to add crops to the cropping pattern, you should review the irrigated area per crop. The shown month (or other interval) is the first month during which the irrigable area is overcharged. Following any corrections, the warning may return for another month.

6.9.1.3 Number of Selected Crops

CRIWAR can calculate the irrigation water requirements of a cropping pattern containing 40 crops. If the 41st crop is selected, CRIWAR gives a warning.

6.9.1.4 Sun Shine Hours

Although, during data entry, CRIWAR has checked the number of daily sunshine hours ($n < 24$ h/day), CRIWAR calculates and checks whether the daily number of sunshine hours, n, does not exceed the maximum possible number of sunshine hours, N, at the related latitude.

References

Allen, R.G. 1996a. Assessing integrity of weather data for use in reference evapotranspiration estimation. *J. Irrig. Drain. Eng.*, ASCE 122(2):97–106.

Allen, R.G. and F.N. Gichuki. 1989. Effects of projected CO_2-induced climate changes on irrigation water requirements in the Great Plains states (Texas, Oklahoma, Kansas and Nebraska). In: The Potential Effects of Global Climate Change on the United States: Appendix C-Agriculture. EPA-230-05-89-053. U.S. Environmental Protection Agency, Office of Policy, Planning and Evaluation, Washington, DC.

Allen, R.G. and C.W. Robison. 2007a. *Evapotranspiration and Irrigation Water Requirements for Crops in Idaho.* Research Completion Report. University of Idaho, submitted to Idaho Department of Water Resources, 292 p.

Allen, R.G. and C.W. Robison. 2007b. Evapotranspiration and Irrigation Water Requirements for Crops in Idaho. USCID 4th International Conference on Irrigation and Drainage, Sacramento, CA, 18 p.

Allen, R.G. and J.L. Wright. 1997. Translating wind measurements from weather stations to agricultural crops. *J. Hydrolog. Eng.*, ASCE 2(1):26–35.

Allen, R.G., M.E. Jensen, J.L. Wright, and R.D. Burman. 1989. Operational estimates of reference evapotranspiration. *Agron. J.*, 81:650–662.

Allen, R.G., L.S. Pereira, D. Raes, and M. Smith. 1998. Crop Evapotranspiration: Guidelines for Computing Crop Water Requirements. Irrigation and Drainage Paper 56. United Nations FAO, Rome, Italy, 300 p. http://www.fao.org/docrep/X0490E/X0490E00.htm

Allen, R.G., W.O. Pruitt, J. Wright, T.A. Howell, F. Ventura, R. Snyder, D. Itenfisu, P. Steduto, J. Berengena, J. Baselga Yrisarry, M. Smith, L.S. Pereira, D. Raes, A. Perrier, I. Alves, I. Walter, and R. Elliott. 2005a. A recommendation on standardized surface resistance for hourly calculation of reference *ET0* by the FAO56 Penman-Monteith method. *Agricultural Water Management*. 81:1–22

Allen, R.G., I.A. Walter, R.L. Elliott, T.A. Howell, D. Itenfisu, M.E. Jensen, and R.L. Snyder. 2005b. *The ASCE Standardized Reference Evapotranspiration Equation.* Report by the Task Committee on Standardization of Reference Evapotranspiration. ASCE, 0-7844-0805-X, 204 pp.

Allen, R.G., L.S. Pereira, M. Smith, D. Raes, and J.L. Wright. 2005c. FAO-56 dual crop coefficient method for estimating evaporation from soil and application extensions. *J. Irrig. Drain. Eng.*, ASCE 131(1):2–13.

Allen, R.G., W.O. Pruitt, D. Raes, M. Smith, and L.S. Pereira. 2005d. Estimating evaporation from bare soil and the crop coefficient for the initial period using common soils information. *J. Irrig. Drain. Eng.*, ASCE 131(1):14–23.

Allen, R.G., M. Smith, A. Perrier, and L.S. Pereira. 1994. An Update for the Definition of Reference Evapotranspiration. *ICID Bulletin*. 43(2):1–34.

Allen, R.G., J.L. Wright, W.O. Pruitt, and L.S. Pereira. 2007. Water requirements. Chapter 8 in Design and Operation of Farm Irrigation Systems, (G.J. Hoffman, R.G. Evans, M.E. Jenser,

D.L. Martin and R.L. Elliott (ed.), 2nd ed., American Society of Agricultural Engineers, MI, P. 208–297.

ASCE-EWRI. 2005. *The ASCE Standardized Reference Evapotranspiration Equation*. Report by the Task Committee on Standardization of Reference Evapotranspiration. ASCE, 0-7844-0805-X, 204 pp.

Ayars, J.E., R.S. Johnson, C.J. Phene, T.J. Trout, D.A. Clark, and R.M. Mead. 2003. Water use by drip-irrigated late-season peaches. *Irrig. Sci.*, 22:187–194.

Bandara, K.M.P.S., 2006, Assessing irrigation performance by using remote sensing, Doctoral thesis Wageningen University, 7 June 2006 (ITC Dissertation number 134), Promoters M.G. Bos and R.A. Feddes, pp 156.

Bastiaanssen, W.G.M., R.A.L. Brito, M.G. Bos, R. Souza, E.B. Cavalcanti, and M.M. Bakker. 2001. Low cost satellite data applied to performance monitoring of the Nilo Coelho Irrigation Scheme, Brazil. Irrigation and Drainage Systems, Kluwer, Dordrecht, Vol. 15.1, pp. 53–79.

Battikhi, A.M. and R.W. Hill. 1986a. Irrigation scheduling and watermelon yield model for the Jordan Valley. *J. Agron. Crop Sci.*, 157:145–155.

Battikhi, A.M. and R.W. Hill. 1986b. Irrigation scheduling and cantaloupe yield model for the Jordan Valley. *Agr. Water Manage.*, 15:177–187.

Belmonte, A.C., A.M. Jochem, A.C. García, A.M. Rodríguez and P.L. Fuster 2006. Irrigation management from space: Towards user-friendly products. *Irrigation and Drainage Systems*, Springer, Dordrecht, Vol.19, 3–4, pp. 337–354.

Blanc, M.L., H. Geslin, I.A. Holzberg, and B. Mason. 1963. Protection against frost damage. Technical Note No. 51. World Meteorological Organisation, Geneva, 62 p.

Blaney, H.F. and W.D. Criddle 1950. Determining Water Requirements in Irrigated Areas from Climatological and Irrigation Data. USDA Soil Cons. Serv. SCS-TP 96. Washington, D.C. 44p.

Bonachela, S., F. Orgaz, F.J. Villalobos, and E. Fereres. 2001. Soil evaporation from drip-irrigated olive orchards. *Irrigation Sci.*, 20(2):65–71.

Bos, M.G. (Ed.) 1976. Discharge Measurement Structures. Ist Edition 1976; 2nd Edition 1978; 3rd Revised Edition 1989. Publication 20. International Institute for Land Reclamation and Improvement/ILRI, wageningen. pp. 399.

Bos, M.G. 1980. Irrigation efficiency at crop production level. *ICID Bulletin*, 29(2):18–26. ICID, New Delhi, July 1980 (also in French and Spanish).

Bos, M.G. 1984. Where water leaves the irrigation system. Deutscher Verband fur Wasserwirtschaft und Kulturbau, *Fortbildungslehrgang Bewasserung*. Bonn. Paper 5, pp. 13. Also **in**: ILRI's Annual Report 1984, pp. 30–38.

Bos, M.G. 1988, Crop irrigation water requirements, ILRI, Wageningen (limited distribution).

Bos, M.G. 1997. Performance indicators for irrigation and drainage. Irrigation and Drainage Systems. Kluwer, Dordrecht, Vol. 11, No. 2, pp. 119–137.

Bos, M.G. 2004. Using the depleted fraction to manage the groundwater table in irrigated areas. Irrigation and Drainage Systems, Kluwer, Dordrecht, Vol. 18.3, pp. 201–209.

Bos, M.G. and J. Nugteren. 1974. On Irrigation Efficiencies. Publication 19. 1st Edition 1974; 2nd Edition 1978; 3rd Revised Edition 1982; 4th Edition 1990. International Institute for Land Reclamation and Improvement/ILRI, Wageningen, pp. 117. 4th Edition also published in Farsi with IRANCID, Tehran.

Bos, M.G., J. Vos, and R.A. Feddes. 1996. CRIWAR 2.0; A Simulation Model on Crop Irrigation Water Requirements. ILRI Publication 46. International Institute for Land Reclamation and Improvement, Wageningen, pp. 117.

Bos, M.G., M.A. Burton, and D.J. Molden. 2005. Irrigation and Drainage Performance Assessment; Practical Guidelines. CABI, Wallingford, UK, 156 pp.

Brunt, D. 1932. Notes on radiation in the atmosphere: I. *Quart. J. Roy. Meteorol. Soc.*, 58:389–420.

Brunt, D. 1952. Physical and Dynamical Meteorology. 2nd Edition. University Press, Cambridge, 428 p.

Burt, C.M., A. Mutziger, D.J. Howes, and K.H. Solomon. 2002. *Evaporation from Irrigated Land in California*. ITCR Report R02-001. Irrigation Training and Research Center, Cal Poly, San Luis Obispo, pp. 166.

Clemmens, A.J. and M. G. Bos. 1990. Statistical methods for irrigation system water delivery performance evaluation. *Irrigation and Drainage Systems* 4:345–365.

Clemmens, A.J., Wahl, T.L., M.G. Bos, and J.A. Replogle. 2001. Water Measurement with Flumes and Weirs. Publication 58. International Institute for Land Reclamation and Improvement, Wageningen, pp. 382.

Costello, L.R., N.P. Matheny, and J.R. Clark. 2000. Estimating the irrigation water needs of landscape plantings in California: Part 1: The landscape coefficient method. University of California Cooperative Extension and California Department of Water Resources.

Choudhury, B.J., N.U. Ahmed, S.B. Idso, R.J. Reginado and C.S.T. Daughtry 1994. Relations between evaporation coefficients and vegetation indices studies by model simulations. *Remote Sensing of Environment* Vol. 50, pp. 1–17.

de Azevedo, P.V., B.B. da Silva, and V.P.R. da Silva. 2003. Water requirements of irrigated mango orchards in northeast Brazil. *Agr. Water Manage.*, 58:241–254.

Doorenbos, J. and A.H. Kassam. 1979. Yield Response to Water. Irrigation and Drainage Paper No. 33 (rev.). FAO, Rome, Italy, 193 pp.

Doorenbos, J. and W.O. Pruitt. 1977. Guidelines for Predicting Crop Water Requirements. Irrigation and Drainage Paper 24. 2nd Edition. FAO, Rome, 156 p.

Droogers, P. and R.G. Allen. 2002. Estimating reference evapotranspiration under inaccurate data conditions. Irrigation and Drainage Systems. Kluwer, Dordrecht, Vol. 16, No. 1, pp. 33–45.

Droogers, P., D. Seckler, and I. Makin, 2001. Estimating the Potential of Rain-Fed Agriculture. IWMI Working Paper 20. Colombo, Sri Lanka, 14 pp.

Everson, D.O., M. Faubion, and D.E. Amos. 1978. Freezing temperatures and growing seasons in Idaho. University of Idaho Agricultural Experiment Station Bulletin 494, 18 p.

Feddes, R.A., P. Kabat, P.J.T. van Bakel, J.J.B. Bronswijk, and J. Halbertsma. 1988. Modelling soil water dynamics in the unsaturated zone: State of the art. *J. Hydrol.*, 100.1/3:69–111.

Girona, J., J. Marsal, M. Mata, and J. del Campo. 2004. Pear Crop Coefficients Obtained in a Large Weighing Lysimeter. ISHS Acta Horticulturae 664. IV 664:277–281

Girona, J., M. Gelly, M. Mata, A. Arbones, J. Rufat, and J. Marsal. 2005. Peach tree response to single and combined deficit irrigation regimes in deep soils. *Agr. Water Manage.*, 72:97–108.

Grattan, S.R., W. Bowers, A. Dong, R.L. Snyder, J.J. Carroll, and W. George. 1998. New crop coefficients estimate water use of vegetables, row crops. *Calif. Agr.* 52(1):16–21.

Grebet, P. and R.H. Cuenca. 1991. History of lysimeter design and effects of environmental disturbances. In: Allen, R.G., Howell, T.A., Pruitt, W.O., Walter, I.A., and Jensen, M.E. (eds.). Lysimeters for Evapotranspiration and Environmental Measurements. ASCE, New York, pp. 10–18.

Haddadin, S.H. and I. Ghawi. 1983. Effect of plastic mulches on soil water conservation and soil temperature in field grown tomato in the Jordan Valley. *Dirasat*, 13(8):25–34.

Hanks, R.J., H.R. Gardner, and R.L. Florian. 1969. Plant growth – evapotranspiration relations for several crops in the Central great Plains. *Agron. J.*, 61.1:30–34.

Hargreaves, G.H. 1994. Defining and using reference evapotranspiration. *J. Irrig. Drain. Eng.*, ASCE 120.6:1132–1139.

Hargreaves, G.H. and R.G. Allen. 2003. History and evaluation of the Hargreaves evapotranspiration equation. *J. Irrig. Drain. Eng.*, ASCE 129(1):53–63.

Hargreaves, G.H. and Z.A. Samani. 1982. Estimating potential evapotranspiration. Technical Note. *J. Irrig. Drain. Eng.*, ASCE 108(3):225–230.

Hargreaves, G.L., G.H. Hargreaves, and J.P. Riley. 1985. Agricultural benefits for Senegal River Basin. *J. Irrig. Drain. Eng.*, ASCE 111:113–124.

Hawkins, R.H., A.T. Hjelmfelt, and A.W. Zevenbergen. 1985. Runoff probability, storm depth, and curve numbers. *J. Irrig. Drain. Eng.*, ASCE 111(4):330–340.

Heilman, J.L., W.E. Heilman and D.G. Moore 1982. Evaluating the crop coefficient using spectral reflectance. Agricultural Journal Vol. 74: pp 967–971.

Hernandez-Suarez, M. 1988. Modeling irrigation scheduling and its components and optimization of water delivery scheduling with dynamic programming and stochastic ET_0 data. Ph.D. dissertation. University of California, Davis, CA.

ICID Committee on Assembling Irrigation Efficiency Data. 1978. M.G. Bos (Chmn.) Standards for the calculation of irrigation efficiencies. *ICID Bulletin* 27(1):91–101. New Delhi (also published in French, Spanish, Turkish, Arabic, and Persian).

Irmak, S., T.A. Howell, R.G. Allen, J.O. Payero, and D.L. Martin. 2005. Standardized ASCE Penman-Monteith: Impact of sum-of-hourly vs. 24-hour time step computations at reference weather station sites. *Trans. ASAE*, 48(3):1063–1077.

Irrigation Association. 2005. *Landscape Irrigation Scheduling and Water Management – Practices Guidelines*. September 2003. Report by Water Management Committee (McCabe, J., Ossa, J., Allen, R.G., Carlton, B., Carruthers, B., Corcos, C., Howell, T.A., Marlow, R., Mecham, B., Spofford, T.L., eds.). 188 p. http://www.irrigation.org/PDF/IA_LIS_AND_WM_SEPT_2003_DRAFT.pdf

Itenfisu, D., R.L. Elliott, R.G. Allen, I.A. Walter. 2003. Comparison of reference evapotranspiration calculations as a part of the ASCE Standardization Effort. *J. Irrig. Drain. Eng.*, ASCE 129(6):440–448.

Jensen, M.E., R.D. Burman, R.G. Allen (eds.). 1990. Evapotranspiration and Irrigation Water Requirements. American Society of Civil Engineering Manual No. 70, 332 p.

Jensen, M.E. and H.R. Haise 1963. Estimating Evapotranspiration from Solar Radiation. J. Irrig. and Drain. Div., ASCE 96, pp. 25–28.

Jensen, M.E., R.G. Allen, T.A. Howell, R.L. Snyder, D.L. Martin, I. Walter. 2007. Evapotranspiration and Irrigation Water Requirements. American Society of Civil Engineering Manual No. 70. 2nd Edition.

Johnson, R.S., L.E. Williams, J.E. Ayars, and T.J. Trout. 2005. Weighing lysimeters aid study of water relations in tree and vine crops. *Calif. Agric.*, 59(2):133–136.

Kabat, P. and J. Beekma. 1994. Water in the Unsaturated Zone. In: Ritzema H.P. (ed.). Drainage Principles and Applications. Publication 16. International Institute for Land reclamation and Improvement, Wageningen, pp. 383–434.

Keller, J. and R.D. Bliesner. 1990. Sprinkle and Trickle Irrigation. van Nostrand Reinhold, New York, 652 p.

Kjelgren, R., L. Rupp, and D. Kilgren. 2000. Water conservation in urban landscapes. *Hort. Sci.*, 35(66):1037–1040.

Kopec, A.R., M.N. Langley, and M.G. Bos. 1984. Major variables which influence effective precipitation. *ICID Bulletin*, 33(2):65–71.

Kruse, E.G. and R.H. Haise. 1974. *Water Use by Native Grasses in High Altitude Colorado Meadows*. Report ARS-W-6-1974. U.S. Department of Agricultural Research Service, Western Region, 60 p.

Ley, T.W., R.G. Allen, and R.W. Hill. 1996. Weather Station Siting Effects on Reference Evapotranspiration. Proceedings of ASAE International Conference on Evapotranspiration and Irrigation Scheduling, San Antonio, TX, pp. 727–734.

Liu, Y., L.S. Pereira, and R.M. Fernando. 2006. Fluxes through the bottom boundary of the root zone in silty soils: Parametric approaches to estimate groundwater contribution and percolation. *Agr. Water Manage*. 84:27–40.

Martin, D. and J. Gilley. 1993. Irrigation Water Requirements. Chapter 2 of Part 623 of the National Engineering Handbook, Natural Resources Conservation Service.

McDonald, M.G. and Harbaugh, A.W. 1988. A modular three-dimensional finite-difference ground-water flow model. Book 6, Chapter A1. U.S. Geological Survey Techniques of Water-Resources Investigations, 586 pp. http://water.usgs.gov/software/modflow.html

Molden, D. J., and T.K. Gates. 1990. Performance Measures for Evaluation of Irrigation Water Delivery System. ASCE Journal of Irrigation and Drainage Engineering, Vol. 116, No 6.

Molden, D.J. 1997. Accounting for Water Use and Productivity. SWIM Paper 1. International Water Management Institute (IWMI), Colombo, Sri Lanka, 16 pp.

Monteith, J.L. 1965. Evaporation and the environment. In: Fogg, G.E. (ed.). The State and Movement of Water in Living Organisms. Cambridge University Press, Cambridge, pp. 205–234.

Neale, C.M.U., W.C. Bausch, and D.E. Heermann. 1989. Development of reflectance based crop coefficients for corn. *Trans. ASAE*, 32(6):1891–1899.

Oldeman, L.R. and M. Frere. 1982. *Technical Report on a Study of the Agro-Climatology of the Humid Tropics of Southeast Asia*. FAO, Rome.

Paço, T.A., M.I. Ferreira, and N. Conceição. 2006. Peach orchard evapotranspiration in a sandy soil: Comparison between eddy covariance measurements and estimates by the FAO 56 approach. *Agr. Water Manage.*, 85(3):305–313.

Pastor, M. and F. Orgaz. 1994. Los programas de recorte de riego en olivar. Agricultura No. 746:768–776 (in Spanish).

Penman, H.L. 1948. Natural Evaporation from Open Water, Bare Soil, and Grass. Proceedings, Royal Society, London, Series A, Vol. 193, pp. 120–146.

Pereira L.S. and I. Alves. 2005. Crop water requirements. In: Hillel, D. (ed.). Encyclopedia of Soils in the Environment. Elsevier, London/New York, Vol. 1, pp. 322–334.

Pereira, L.S., A. Perrier, R.G. Allen, and I. Alves. 1999. Evapotranspiration: Concepts and future trends. *J. Irrig. Drain. Eng.*, ASCE 125(2):45–51.

Pittenger, D.R. and J.M. Henry. 2005. Refinement of Urban Landscape Water Requirements. University of California Cooperative Extension, South and Central Coast and South Region.

Pittenger, D.R. and D. Shaw. 2001. Applications of Recent Research in Landscape Irrigation Management. Proceedings of UCR Turfgrass and Landscape Management Field Day, Riverside, USA, spp. 17–18.

Pittenger, D.R., D.A. Shaw, D.R. Hodel, and D.B. Holt. 2001. Responses of landscape groundcovers to minimum irrigation. *J. Environ. Hort.*, 19(2):78–84.

Pruitt, W.O. 1986. Traditional Methods: 'Evapotranspiration Research Priorities for the Next Decade'. ASAE Paper No. 86-2629, 23 p.

Pruitt, W.O., E. Fereres, P.E. Martin, H. Singh, D.W. Henderson, R.M. Hagan, E. Tarantino, and B. Chandio. 1984. Microclimate, Evapotranspiration, and Water-Use Efficiency for Drip- and Furrow-Irrigated Tomatoes. Proceedings of 12th Congress, International Commission on Irrigation and Drainage, Ft. Collins, CO, pp. 367–394.

Raes, D. (2004). UPFLOW. Water Movement in a Soil Profile from a Shallow Water Table to the Topsoil (capillary rise). Reference Manual Version 2.2. K.U. Leuven University, Faculty of Applied Bioscience and Engineering, Department Land Management, Vital De Costerstraat 102, 3000 Leuven, Belgium, 18 p.

Richie, W.E. and D.R. Pittenger. 2000. Mixed Landscape Irrigation Research Findings. Proceedings of UCR Turfgrass and Landscape Management Research Conference and Field Day, Riverside, USA pp. 12–13.

Rogers, J.S., L.H. Allen, and D.J. Calvert. 1983. Evapotranspiration for humid regions: Developing citrus grove, grass cover. *Trans. ASAE*, 26(6):1778–1783, 1792.

Safadi, A.S. 1991. Squash and cucumber yield and water use models. Unpublished Ph.D. dissertation, Department of Biological and Irrigation Engineering, Utah State University, Logan, UT 84322-4105, 190 p.

Senga, Y. and J.F. Mistry 1989. Field application efficiency termonology for paddy rice. *ICID Bulletin*, New delhi, Vol. 38 (1), pp 61–70.

Shaw, D.A. and D.R. Pittenger. 2004. Performance of landscape ornamentals given irrigation treatments based on reference evapotranspiration. *Acta Hort.* 664:607–613.

Smith M. 1900. Draft Report on the Expert Consultation on Revision of FAO Methodologies for Crop Water Requirements. FAO, Rome. 45 pp.

Smith, M., R.G. Allen, J.L. Monteith, A. Perrier, L.S. Pereira, and A. Segeren. 1991. Report of the Expert Consultation on Procedures for Revision of FAO Guidelines for Prediction of Crop Water Requirements. UN-FAO, Rome, Italy, 54 p.

Smith, M., R.G. Allen, and L.S. Pereira. 1996. Revised FAO Methodology for Crop Water Requirements. Proceedings of International Conference on Evapotranspiration and Irrigation Scheduling, ASAE, San Antonio, TX, pp. 116–123.

Snyder, R. L. and S. Eching. 2004. Landscape Irrigation Management Program—IS005 Quick Answer. University of California, Davis, CA. http://www.biomet.ucdavis.edu/irrigation_scheduling/LIMP/limp.pdf.

Snyder, R.L. and S. Eching. 2005. Urban Landscape Evapotranspiration. California State Water Plan, Sacramento, CA, Vol. 4, pp. 691–693. http://www.waterplan.water.ca.gov/reference/index.cfm#infrastructure

Snyder, R.L., B.J. Lanini, D.A. Shaw, and W.O. Pruitt. 1989a. Using Reference Evapotranspiration (ET_0) and Crop Coefficients to Estimate Crop Evapotranspiration (ETc) for Agronomic Crops, Grasses, and Vegetable Crops. Cooperative Extension, University of California, Berkeley, CA, Leaflet No. 21427, 12 p.

Snyder, R.L., B.J. Lanini, D.A. Shaw, and W.O. Pruitt. 1989b. Using Reference Evapotranspiration (ET_0) and Crop Coefficients to Estimate Crop Evapotranspiration (ETc) for Trees and Vines. Cooperative Extension, University of California, Berkeley, CA, Leaflet No. 21428, 8 p.

Snyder, R.L., J.P. De Melo-Abreu, and S. Matulich. 2005. Frost Protection: Fundamentals, Practice and Economics. FAO Environment and Natural Resources Service Series 10. FAO, Rome, Vol. 2, 240 pp and 64 pp + CD-ROM. Soil Conservation Service. 1982. National Engineering Handbook. U.S. Printing Office.

Tasumi, M., R.G. Allen, R. Trezza, J.L. Wright. 2005. Satellite-based energy balance to assess within-population variance of crop coefficient curves, *J. Irrig. Drain. Eng.*, ASCE 131(1):94–109.

Testi, L., F.J. Villalobos, and F. Orgaza. 2004. Evapotranspiration of a young irrigated olive orchard in southern Spain. *Agr. Forest Meteorol.*, 121(1,2):1–18.

Till, M.R., and M.G. Bos. 1985. The Influence of Uniformity and Leaching on the Field Application Efficiency, *ICID Bulletin*, New Delhi, Vol. 34(1), pp 32–36.

Thornton, P.E. and S.W. Running. 1999. An improved algorithm for estimating incident daily solar radiation from measurements of temperature, humidity, and precipitation. *Agr. Forest Meteorol.*, 93:211–228.

Turc, L. 1954. Le bilan d'eau des sols. Relations entre les précipitations, l'évaporation et l'écoulement. Ann. Agron. 6, pp. 5–131.

USDA Soil Conservation Service. 1970. Irrigation Water Requirements. Technical Release 21.

USDA Soil Conservation Service. 1972. USDA-SCS National Engineering Handbook, Section 4, Table 10.1.

Van der Kimpen, P.J. 1991. Estimation of crop evapotranspiration by means of the Penman-Monteith equation. Unpublished Ph.D. dissertation, Department of Biological and Irrigation Engineering, Utah State University, Logan, UT 84322-4105, 242 p.

Ventura, F., D. Spano, P. Duce, and R.L. Snyder. 1999. An evaluation of common evapotranspiration equations. *Irrig. Sci.*, 18:163–170.

Villalobos, F.J., F. Orgaz, L. Testi, and E. Fereres. 2000. Measurement and modeling of evapotranspiration of olive orchards. *Eur. J. Agron.*, 13:155–163.

Wesseling, J.G. 1991. *CAPSEV: Steady-State Moisture Flow Theory, Program Description and User Manual*. Report 37. Winand Staring Centre, Wageningen, 51 p.

World Bank. 1999. World Bank Atlas from the World Development Indicators. World Bank, Washington, DC, pp. 63.

Wösten, J.H.M. 1987. Beschrijving van de waterretentie- en doorlatenheids-karakteristieken uit de Staringreeks met analytische functies. Rapport, Stichting voor Bodemkartering, 2019, Wageningen, 54 pp. (in Dutch).

Wright, J.L. Dec 1981. Crop coefficients for estimates of daily crop evapotranspiration. Irrigation Scheduling for Water and Energy Conservation in the 80's. American Society of Agricultural Engineering, MI, Mpp. 18–26.

Wright, J.L. 1982. New evapotranspiration crop coefficients. *J. Irrig. Drain. Div.*, ASCE St. Joseph, USA108:57–74.

Wright, J.L. 1990. Evapotranspiration data for dry, edible beans, sugar beets, and sweet corn at Kimberly, Idaho. Unpublished data, USDA-ARS, Kimberly, ID.

Index